Toward Efficient Military Retirement Accrual Charges

James Hosek, Beth J. Asch, Michael G. Mattock

Prepared for the United States Army

For more information on this publication, visit www.rand.org/t/rr1373

Library of Congress Cataloging-in-Publication Data is available for this publication.

ISBN: 978-0-8330-9295-3

Published by the RAND Corporation, Santa Monica, Calif.

The RAND Corporation is a research organization that develops solutions to public policy challenges to help make communities throughout the world safer and more secure, healthier and more prosperous. RAND is nonprofit, nonpartisan, and committed to the public interest.

RAND's publications do not necessarily reflect the opinions of its research clients and sponsors.

Preface

This report provides results from a research project titled "Efficient Retirement Accrual Charges." The purpose of the project was to develop a principled and evidence-based case for reforming the retirement accrual charge system.

This report develops a theoretical model explaining why the current accrual system produces inaccurate estimates of each service's accrual charges, reviews past critiques of the system, and presents empirical estimates of the bias in each service's accrual charges under the current system as compared with a service-specific system. The research and findings are likely to be of interest to policy communities with responsibility for force mix decisions with respect to personnel experience and capital/labor trade-offs, and for the efficient allocation of resources in programming, budgeting, and operations.

This research was sponsored by the Office of the Assistant Secretary of the Army (Financial Management and Comptroller) and conducted within the RAND Arroyo Center's Personnel, Training, and Health Program. RAND Arroyo Center, part of the RAND Corporation, is a federally funded research and development center sponsored by the United States Army.

The Project Unique Identification Code (PUIC) for the project that produced this document is HQD146689.

Contents

Figures and Tables

Figures

Tables

Summary

The retirement accrual charge system introduced in 1984 brought visibility to the military retirement liability resulting from personnel policy decisions to ensure that future military retirement benefits would be funded. However, because it mandates a single accrual charge rate for all services, the system sends inaccurate signals of the total and marginal costs of the accruing liability to decisionmakers. This makes the total costs of the liability accruing to the Army, Navy, and Marine Corps active components (ACs) look higher than they are in reality, and the cost of the liability accruing to the Air Force AC looks lower. The single accrual charge rate also tends to distort marginal costs by diluting the effect of any one service's change in force shape so that the service does not realize the full consequences of its own incremental personnel decisions that affect experience mix. Removing the inaccuracy to improve the information for decisionmakers could help to enhance resource allocation efficiency among the services and, within services, improve the efficiency of personnel decisions and budget choices between personnel and other resources. Gains from efficiency could be used to improve national defense or decrease its cost.

This document presents a theoretical model explaining why the single accrual charge rate produces inaccurate estimates of each service's total accrual charge, i.e., charges that are higher, or lower, than a service's actual accruing retirement liability. The model also shows why the single accrual charge rate makes a service's marginal accrual charge inaccurate, yet this charge is relevant to decisions regarding personnel versus non-personnel resources and to decisions regarding the experience mix of personnel. In particular, we develop expressions to show the inaccuracy in the total and marginal accrual costs, in which the marginal cost refers to the change in total accrual cost resulting from an incremental change in force size or the experience mix of personnel.

The document also reviews past critiques of the single accrual rate and considers their proposals and recommendations for change. Finally, the document presents empirical estimates of the inaccuracies in each service's total and marginal accrual charges under the current system of a single accrual charge rate versus a service-specific rate. The estimates are done both for the current military retirement system and for a reformed retirement system.

We find that the Army's total annual retirement accrual charge would be approximately $400 million less under a service-specific accrual charge rate. The charge also would be lower for the Navy and Marine Corps but higher for the Air Force. With respect to the marginal retirement accrual cost, it depends in general on the policy under consideration. In a specific example we pursue, a decrease in Army force size with a drop in average seniority would reduce the accrual charge by $200 million under the single accrual charge rate but $361 million under a service-specific accrual charge rate. Similarly, the single rate would result in too low an increase in accrual cost from a reverse example, i.e., an increase in force size that also increased average seniority.

We find that the total and marginal accrual costs under the Military Compensation and Retirement Modernization Commission's proposed retirement reform—a particular policy alternative we consider—would behave in a fashion similar to the total and marginal accrual costs under the current retirement system. Under the proposed retirement reform, the Army's total annual accrual charge would be approximately $380 million less under a service-specific accrual charge rate than under the single accrual system. As in the current retirement system, the charge would be lower for the Navy and Marine Corps but higher for the Air Force. The inaccuracy in marginal accrual cost would still be present, although somewhat reduced because of the reduction in retirement liability.

Thus, today's single accrual charge rate makes the annual accrual cost of three of the four services look too high, and makes one service's cost look too low. The single rate also biases the marginal cost of policy actions, depriving decisionmakers of the necessary accurate cost information to make efficient decisions. We conclude with a discussion on issues related to implementing a service-specific accrual rate in place of the current single rate.

Acknowledgments

We appreciate the support and guidance offered by Tim Bonds, Vice President, RAND Arroyo Center, and Michael Hansen, Director, Personnel, Training, and Health Program, RAND Arroyo Center. We are grateful to LTG (retired) Joseph E. Martz, former Military Deputy for Budget, Office of the Assistant Secretary of the Army for Financial Management and Comptroller (ASA [FM&C]) for his office's sponsorship of this research and helpful comments on our work in progress. We thank Saul Pleeter for providing a history of the current policy. Our reported benefited from reviews provided by Al Robbert of RAND and David Rodney of the CNA Corporation.

CHAPTER ONE

Introduction

Primary elements of the costs of military personnel include pay and benefits during military service, and retirement and health care benefits paid to qualified personnel after leaving the military. This report focuses on retirement benefits. Public Law (PL) 98–94, enacted in 1984 and implemented in fiscal year (FY) 1985, mandated accrual accounting to fund the military retirement benefit liability and specified the use of the aggregate entry-age normal accounting method. Before PL 98–94, the amount appearing in the Department of Defense (DoD) budget for military retirement was the annual payment to current military retirees; military retirement was a pay-as-you-go (PAYGO) system. The shift to accrual accounting sought to meet the objective of recognizing in the current budget the cost of future retirement benefits associated with current manning decisions.

Under PL 98–94, DoD includes an accrual charge based on aggregate entry-age normal costing in the budget transmitted to Congress.[1] Paid year after year and estimated by the DoD Office of the Actuary, accrual charges can be expected to fully fund the retirement liability of entering cohorts of military personnel, taking into account

[1] According to 10 U.S. Code, Section 1465–Determination of Contributions to the Fund,

> "The amount determined under paragraph (1) [the annual accrual charge] for any fiscal year is the amount needed to be appropriated to the Department of Defense for that fiscal year for payments to be made to the Fund during that year under Section 1466 (a) of this title. The President shall include not less than the full amount so determined in the budget transmitted to Congress for that fiscal year under Section 1105 of Title 31. The President may comment and make recommendations concerning any such amount."

the expected growth in military pay, interest rates, and other factors such as mortality rates. The accrual charge is computed as the product of the normal cost percentage (NCP) and the annual basic pay bill. The NCP is the level percentage of basic pay that must be contributed over the entire military career of a group of new entrants to pay for that group's future retirement and survivor benefits. By law, there are separate NCPs for full-time (e.g., active component [AC]) and part-time (Selected Reserve) military personnel. The NCPs do not vary by service or any other characteristic, such as years of service.

The move to accrual accounting was intended to improve manpower management by including a measure of future retired pay costs alongside current personnel costs when considering current force structure decisions (Congressional Budget Office [CBO], 1983). Under PAYGO, increasing or decreasing the size of the current force has no effect on retirement benefits paid to currently retired personnel and hence no effect on current outlays for retirement benefits; however, under an accrual accounting system, retirement costs would increase. Similarly, the CBO argued that accrual accounting would ensure that DoD and Congress faced the full cost of pay raise decisions, given that retirement benefits depend on the pay raise. CBO added that accrual charges would be sensitive to technical assumptions about future pay raises, interest rates, and other costs. Therefore, it recommended the establishment of an independent board of actuaries to determine the appropriate technical assumptions; so PL 98–94 established the DoD Office of the Actuary.

A number of studies and commissions argue that PL 98–94 is flawed and recommend reforms. Among the key concerns is the use of a single NCP for all personnel, given full-time or part-time status. A single NCP generates accrual charges that are inaccurate for each military service. As a result, the military services do not have accurate cost information for resource allocation decisions. Further complicating the situation, the accrual charge does not accurately capture the retirement cost increase or decrease from alternative experience mixes or changes in military pay, again resulting in inaccurate signals. Because distorted cost signals enter planning, programming, and budgeting processes, the military services may not be able to achieve an efficient allocation of personnel with respect to their experience mix and relative to other resources, such as weapons procurement, training, operations, and more. Past studies have offered different recommendations on how to address these distortions, including using service-specific NCPs.

Critiques of the single NCP have brought up other issues related to incentives for efficient resource allocation, namely, whether a service can recoup any past overfunding of its accruing retirement benefit liability, and whether a service's current accrual charge can be decreased in anticipation of decreased future liabilities resulting from current service plans.

Reforms to the military retirement system are under consideration. While these discussions have not included reforms to the way the accrual charge is computed, the efficiency of the retirement accrual charge system is highly relevant to how retirement reform will impact the cost of changes to force size and experience mix.

The research summarized in this report analyzes the efficiency of the retirement accrual charge system. Specifically, it develops a principled and evidenced-based case for reforming the system, arguing that a system that sets the NCP by service and separately for enlisted and officers within service would improve the efficiency of the system. Our assessment of the accrual system is based on theoretical and quantitative methods and builds on past studies that have considered the efficiency of the system.

In Chapter Two, we develop a heuristic model that enables us to consider the inaccuracies in total and marginal personnel costs associated with the single NCP approach mandated by law. In addition, we consider how policy actions that change the size and/or experience mix of the force affect cost and how the single NCP approach results in inaccurate cost signals associated with these policy changes. In particular, we develop expressions to show the difference in the total accrual cost and in the marginal accrual cost under single versus service-specific NCPs, where the marginal cost refers to the change in total accrual cost resulting from an incremental change in force size or in the experience mix of personnel. This discussion is a prelude to Chapter Four, where empirical estimates of these biases are presented. Furthermore, the estimates in Chapter Four are done both for the current military retirement system and a reformed retirement system.

Second, we draw the key insights of past studies and review the main recommendations for changing the current approach for funding the military retirement system. One drawback of past studies is that they do not provide any estimates of the total or marginal cost to the military services of the accrual charge nor cost savings to the services of changing the accrual system. That is, they do not provide quantitative evidence on the size of the inaccuracy in total and marginal costs associated with using a single NCP approach. Thus, our approach also includes estimating the accrual charge to each service, separately for officers and enlisted personnel, under a single versus service-specific system. It also includes estimating the marginal cost of the accrual charge to the services under a single versus service-specific system and how these estimates change with force size and/or experience mix. Thus, we provide quantitative estimates of the inaccuracies in total and marginal accrual costs under a single versus service-specific NCP. Furthermore, to illustrate how retirement reform would affect this, we also compute estimates under the Military Compensation and Retirement Modernization Commission's recent retirement reform recommendation (MCRMC, 2015).

We use the RAND Corporation's dynamic retention model (DRM) for all of the accrual charge computations. It is important to note that the DoD Office of the Actuary performs all official computations of the accrual charge. Our estimates are not intended to replace those estimates, and indeed, should policymakers choose to reform the accrual system, the DoD Office of the Actuary would and should provide estimates of how the accrual charges would change. Instead, our estimates are intended to inform the policy debate by showing the magnitude of the inaccuracies in the current accrual charge method.

Organization of This Document

Chapter Two presents the heuristic model we developed for understanding the NCP and the cost signals it provides for resource allocation. Chapter Three describes insights and recommendations from past studies of the military accrual system and discusses challenges for reform. Chapter Four provides estimates of Army personnel cost changes, as well as changes in costs to the other military services in a shift from a single NCP to service-specific NCPs. Chapter Four also contains estimates of the marginal costs of a policy change under a single versus service-specific NCPs, as well as total and marginal accrual costs for a retirement reform example. Chapter Five draws the main conclusions and discusses steps toward implementing service-specific accrual charges.

CHAPTER TWO

NCP and Cost: A Heuristic Model

Introduction

This chapter presents a heuristic model to define single and service-specific NCPs, and uses the model to show expressions for the difference in total and marginal accrual costs from a single NCP as compared with service-specific NCPs. For marginal cost, we consider two types of policy change, both of which can change force size and experience mix: a personnel policy action that changes the retention rate but not military pay; and a change in military pay. We also consider a change in experience mix that holds force size constant. The model is structured around the expected career cost (lifecycle cost) of an entering cohort of personnel and the expected retirement liability of the cohort. The model introduces issues arising when a single NCP is used instead of service-specific NCPs and provides background for the empirical analysis in Chapter Four based on the DRM. Finally, we refer to the model as heuristic because we use it to describe the basic framework of entry-age normal accounting; in doing this, we use a simplifying assumption. The assumption is that the year-to-year continuation rate of military personnel in a specific branch is the same in all years of service. Use of this assumption simplifies the derivations in the heuristic model and, we believe, makes the analysis more intuitive. However, the model used by the DoD Actuary to compute the NCP does not assume that the continuation rate is the same for all years of service but instead allows it to differ by year of service based on historical realizations of the continuation rates. Our empirical estimates of NCPs, presented in Chapter Four, also allow the continuation rate to differ by year of service. Therefore, the assumption used in this chapter is only for ease of exposition of the basic ideas underlying the NCP and the reason for differences between single and service-specific NCPs.

Expected Years of Service from an Entering Cohort

As a step toward the NCP, we need to know the cumulative retention profile. In the model, cumulative retention to a given year of service equals a continuation rate r raised to an exponent of the year of service minus 1. The model assumes the rate is the same for all years of service, a strong assumption. Retention in the first year of service for an entering cohort is $r^0 = 1$, indicating that the probability of being present during that year is one. At the immediate end of year one, or equivalently the immediate start of year two, the fraction r^1 of the cohort is retained, and this fraction of the entering cohort is present during year two. After the completion of 20 years of service, the fraction r^{20} is present, and so on for the following years. Strung together by year of service, these fractions comprise the retention profile. The expected person-years of service over a 30–year career from an entering cohort of size n equals n times the sum of the fractions present during each year of service:

$$n\sum_{y=1}^{30} r^{y-1}.$$

This evaluates to $n/(1 - r)$ when the military career is infinitely long, and the approximation is fairly accurate for a 30–year career.[1] The term $1/(1 - r)$ is a member's expected length of service, i.e., average years of service. We confirm this by computing the expected length of service in the usual way, by taking the sum over years of service of the probability of a career that is y years long and no longer $[r^{y-1}(1 - r)]$, multiplied by y:

[1] For example, when $r = 0.9$, an individual's expected years of service are ten for an infinitely long career and 9.58, or 4.2 percent less, for a 30–year career.

$$\sum_{y} r^{y-1}(1-r)y.$$

As the potential length of a career becomes infinitely long, this expression simplifies to $1/(1-r)$.

Normal Cost Percentage

The NCP is defined as the level (constant) percentage of basic pay over the career of an entering cohort that, when invested, is sufficient to fund the retirement liability of the cohort. The present discounted value in year zero of the basic pay received by members of a cohort over a career is

$$W = n\sum_{y=1}^{30} r^{y-1} b_y \beta^{y-1},$$

where b_y is average basic pay during year y and β is the discount factor. By definition, $\beta = 1/(1 + \delta)$, where δ is the government discount rate or alternatively its rate of return on financial assets.

Assume an amount equal to a given percentage θ of basic pay at each year of service is invested at the beginning of the year in a government retirement fund. The present value of the fund at the completion of 20 years of service is

$$F_{20} = n\sum_{y=1}^{30} r^{y-1}\, \theta\, b_y (1+\delta)^{20-(y-1)} = n\theta\sum_{y=1}^{30} r^{y-1} b_y \beta^{(y-1)-20}.$$

Notice that the expression covers an entire 30–year career.

The retirement liability depends on the probability of retiring in each year after 20 years of service, up to mandatory retirement after 30 years, and beginning to draw retirement benefits at that point. Retirement benefits are paid over the remaining lifetime; for simplicity, we assume the lifetime is infinite (making it infinite does not affect the basic results but makes the expression simpler). The present discounted value of the expected liability for retirement benefits given the completion of 20 years of service is

$$L_{20} = n\{\sum_{y=20}^{29} r^{y-1}(1-r)\frac{1}{\delta}(0.025y\,b_y)\beta^{[(y-1)-20]} + r^{(30-1)}\frac{1}{\delta}[0.025(30)b_{30}]\beta^{[(30-1)-20]}\}.$$

The fraction of the cohort retiring at y years of service is $r^{y-1}(1-r)$. The 0.025 constant is the retirement multiplier, which is 2.5 percent under current policy. Thus, $0.025\,y\,b_y$ is the retirement benefit for a member retiring with y years of service. Assuming this benefit is paid over a lifetime, the value of this benefit stream is the benefit amount times $\beta^{[(y-1)-20]}/\delta$. Here, the factor $1/\delta = \beta/1-\beta$ converts the lifetime retirement benefit stream into its present value assuming the lifetime is infinite. The term $\beta^{[(y-1)-20]}$ brings this present value from year of service y to year of service 20. The second term in brackets applies to those completing 30 years of service, all of who are required to retire (in this expression, $(1-r)$ is forced to equal 1).

Using 20 years of service as the point at which the fund equals the retirement liability is natural given retirement eligibility at 20 years, but it is equally meaningful to compare the fund and liability at time zero. To do this, we multiply the present value at year 20 by the discount factor β^{20}, which gives:

$$F = n\theta\sum_{y=1}^{30} r^{y-1}\,b_y\,\beta^{(y-1)} \equiv \theta W$$

$$L = n\frac{1}{\delta}\{\sum_{y=20}^{29} r^{y-1}(1-r)(0.025\,y\,b_y)\beta^{(y-1)} + r^{(30-1)}[0.025(30)b_{30}]\beta^{(30-1)}\}.$$

From these two expressions and the requirement that the fund equal the liability, the normal cost percentage is

$$\theta = \frac{L}{W},$$

where L and W are respectively the present value at time zero of the cohort's expected retirement liability and expected basic pay over the career. Notice that θ, the NCP, is independent of n holding r constant. Thus, holding the retention profile constant but adjusting the entry cohort—and, hence, force size—up or down does not change the NCP.

As an example of NCP computation, assume basic pay is linear in years of service: $b_y = b_0 + b_1 y$. Supporting this assumption, the CBO finds average basic pay by year of service to be nearly linear for enlisted personnel.[2] The model has not treated enlisted and officers separately, so we choose basic pay parameters to approximate average basic pay for officers and enlisted combined. Average basic pay (officers and enlisted together) in 2014 was about \$22,000 at entry (b_0) and increased to \$60,000 in the 20th year of service, which gives an average increase of \$2,000 per year of service (b_1). A retention rate of 0.91 implies that 18.3 percent of an entering cohort will complete 20 years of service, qualifying to retire from the AC. The government real discount rate in 2012 was 3.15 percent.[3] These parameters produce an NCP of 38.6 percent, which is lower than the DoD Office of the Actuary's value of 43.3 percent in 2013.[4] However, the Actuary's computation also allows for funds contributed to the retirement system for service members who later go on to retire from the reserve component (RC) to be credited toward funding the RC retirement liability. If we assume that the present value of 9 percent of the total funds contributed in years of service one through 19 are subtracted from W and in effect moved to the RC,[5] then these parameters produce an AC NCP of 42.3 percent, which is 1 percentage point (and 2.2 percent) less than the DoD Actuary's

---

[2] See, for example, Economic and Budget Issue Brief from the Congressional Budget Office (2007).

[3] See, for example, the memorandum to file from the DoD Office of the Actuary (2014) on "Actuarial Work for the Chief Financial Officers Act Financial Statements."

[4] See, for example, the DoD Office of the Actuary (2013) report, *Valuation of the Military Retirement System: September 30, 2011.*

[5] The 9 percent is an estimate of the portion of an AC officer/enlisted entering cohort that qualifies for Reserve retirement. These personnel members serve in the Reserves after leaving active duty and accumulate 20 creditable years of service, including years served in the AC. The estimate is based on the authors' tabulations using data on 1990–1991 entering cohorts.

value.[6] We emphasize that this computation is only illustrative, and the actual computation is far more detailed and considers additional factors such as disability retirement, regular versus part-time (Reserve) service, and the military's several but closely related retirement systems. Furthermore, as mentioned, neither the DoD Actuary nor we use a constant continuation rate in computing an NCP.

Effect of a Single NCP on a Service's Total Manpower Cost

Suppose there are two military services, service one and service two. Under a single NCP, the services' basic pay bill and retirement liability will be pooled. Assuming that service two's NCP is higher than service one's NCP, the single NCP will be higher than service one's. The extent to which it is higher depends on service two's share of the present value of career basic pay for the two services.

The service-specific and single NCPs are

[6] The DoD Office of the Actuary reports the NCP consists of two portions. The first can be thought of as representing the NCP before the introduction of legislation permitting concurrent receipt of DoD retirement benefits and Department of Veterans Affairs (VA) disability benefits. This NCP is about 32 percent. The second is the portion of the NCP deriving from the need to fund the retirement liability associated with concurrent receipt. This NCP is about 10 percent. The 43.1 percent in the example above is for the "full" NCP as currently calculated under concurrent receipt. However, Treasury funds the portion of the retirement accrual charge associated with concurrent receipt. So, for instance, the retirement accrual charge in a service's budget would be based on the 32 percent. We keep this distinction in mind in our calculations in Chapter Four where we compute the marginal cost to a service under service-specific versus single NCPs; that is, those calculations are based solely on the portion of the accrual charge paid by the service, not the Treasury's portion.

$$\theta_1 = \frac{L_1}{W_1}$$

$$\theta_2 = \frac{L_2}{W_2}$$

$$\theta = \frac{L_1 + L_2}{W_1 + W_2}.$$

Since $L_1 = \theta_1 W_1$ and $L_2 = \theta_2 W_2$, the single NCP can be written as

$$\theta = \frac{W_1\theta_1 + W_2\theta_2}{W_1 + W_2}.$$

Letting $\alpha = W_1 /(W_1 + W_2)$, we have $\theta = \alpha\, \theta_1 + (1 - \alpha)\, \theta_2$.

The difference between the single and service-specific NCP for service one equals:

$$\theta - \theta_1 = (1 - \alpha)(\theta_2 - \theta_1).$$

Therefore, the difference in the NCPs is larger the greater service two's share of the total basic pay bill and the greater the difference between the NCP for service two and service one. If there were no difference, there would of course be no bias.

A single NCP causes service one's retirement accrual charge to be higher than it should be; service one's own NCP is the correct percentage under entry-age normal accounting to cover its retirement liability. Use of the higher NCP increases service one's accrual charge. Specifically, the cost over the military career of a cohort entering service one is the sum of the present value of its basic pay outlay and retirement liability:

$$C_1 = W_1 + L_1 = W_1(1 + \theta_1).$$

The expression $W_1 + L_1$ is the actual expected cost of the entering cohort over its military lifecycle, and $W_1(1 + \theta_1)$ re-expresses this cost in terms of the NCP.

Under a single NCP, θ replaces θ_1 and service one's expected manpower cost is higher by $W_1(\theta - \theta_1)$. Making service one appear more costly than it is, a single NCP might affect resource allocation decisions when building a budget submission, deciding on budget authorization and appropriation, or executing a budget. There could be pressure on other parts of the personnel budget, such as for training and family support, decreased support for the service's requested end strength, or decreases in budgets for procurement, operations and maintenance, or construction.

A single NCP treats military services as though the average retirement liability per member is the same for every service, but in reality it differs. As shown in Chapter Four, the retirement liability for an entering cohort is highest in the Air Force and lowest in the Marine Corps, with the Army and Navy in between. A single NCP makes the Air Force accrual charge less than it should be to fund its retirement liability, and makes the Marine Corps, Navy, and Army accrual charges too large.

A Single NCP Distorts Marginal Incentives

Proposals frequently involve changes at the margin, e.g., changing force size or experience mix. It is therefore useful to ask how the single NCP affects the marginal cost of a change relative to the marginal cost under a service-specific NCP. The cases we consider are the marginal cost of (a) an increase in force size holding experience mix constant, (b) an increase in experience mix via nonpecuniary actions that change the year-to-year continuation rate r, allowing force size to change, (c) the same but holding force size constant, and (d) an across-the-board increase in basic pay. Changes (a) through (c) do not involve changes in military pay but do alter the overall cost of the force because a larger or more experienced force is more expensive. An increase in basic pay increases budget costs directly (basic pay and retired pay are higher, given that retired pay depends on basic pay) and indirectly through higher retention and consequently a higher basic pay bill and retired pay bill because of more people staying in active service and vesting in retirement. Examples of nonpecuniary policy changes affecting retention include the use of service member preferences in making assignments, less frequent moves, less restrictive retention controls, the provision of training and education transferable to civilian jobs, improved military housing, improved family support, and others.

Although the changes we look at only involve force size, experience mix, and basic pay, biased marginal costs in any of those dimensions can also influence the trade-off between military personnel and military capital. Thus, the discussion of bias in marginal incentives has implications extending to efficient resource allocation in general—the marginal cost of personnel is biased relative to the marginal cost of other defense inputs.

A service's capability in terms of personnel depends on its size and experience mix. These depend on entry cohort size and retention through a notional production function $F(n, r)$ that relates force capability to personnel force size and experience mix. Force size depends on entry cohort size and year-to-year retention r. Experience mix is represented by r because in the model an increase in r causes an increase in average experience, i.e., in experience mix. Furthermore, because force size and experience come at a price, it is relevant to consider capability relative to cost. In Chapter Three, we will argue that because Congress ultimately holds the purse strings, a service cannot unilaterally decide to increase or decrease its size or experience mix. Here, we assume that a service and Congress work together (interact cooperatively) to reach agreement on the best force size and experience mix, recognizing that increasing force size or experience mix can boost military capability but also costs. For service one, suppose the objective is to maximize capability net of cost (though other objectives could be considered such as maximizing capability subject to a given personnel budget, or minimizing cost subject to a given level of capability):

$$F_1(n_1, r_1) - (W_1 + L_1)$$

or equivalently

$$F_1(n_1, r_1) - W_1(1+\theta_1).$$

Consistent with the discussion above, F should be considered the present discounted value of an entering cohort's expected capability, and W and L are respectively the present discounted values of basic pay and retirement liability over the cohort's service career.

Marginal Cost of Increasing Force Size

Holding the continuation rate constant, force size is optimized at the point where the marginal improvement in capability from an increase in entry cohort size equals the marginal the marginal cost of such an increase. This condition is expressed as

$$F_{1n_1} - W_{1n_1}(1+\theta_1) = 0,$$

where the subscript n_1 denotes the partial derivative with respect to entry cohort size for service one.[7]

[7] Because r is being held constant, the NCP does not change when force size increases; as mentioned, the NCP is independent of force size but does depend on experience mix. From above,

$$n\sum_{y=1}^{30} r^{y-1} b_y \beta^{(y-1)} = W,$$

so

$$W_{1n_1} = \sum_{y=1}^{30} r^{y-1} b_y \beta^{(y-1)} = \frac{W}{n}.$$

This implies that the marginal increase in the present discounted value of the cohort's expected basic pay with respect to entry cohort size is equal to the average present discounted value per member. Therefore, the above expression can also be written as

$$F_{1n_1} - \frac{W_1}{n_1}(1+\theta_1) = 0.$$

If a single NCP is used instead of the specific NCP for force one, the marginal cost is $W_{1n_1}(1 + \theta)$. The marginal cost associated with the accrual charge is $\theta_1 W_{1n_1}$ under a service-specific NCP and θW_{1n_1} under a single NCP approach.

The bias in marginal cost caused by a single NCP is the difference between $W_{1n_1}(1 + \theta)$ and $W_{1n_1}(1 + \theta_1)$, or $W_{1n_1}(\theta - \theta_1)$. Looking ahead to Chapter Four, the results from our DRM analysis (Table 4.6) confirm that as expected from this expression the bias is always negative for officers in each of the services—the use of a single NCP causes the marginal cost to appear smaller than it is when calculated under service-specific NCPs—and the bias is negative for Air Force enlisted but positive for Army, Navy, and Marine Corps enlisted.

Marginal Cost of Increasing Experience Mix by Nonpecuniary Means

Increasing the retention rate increases the retention profile. We consider this case as a stepping-stone to the example below, where the retention rate increases in response to an increase in pay. In the present case, with the intercept remaining fixed at 1, the increase in *r* increases cumulative retention to each year of service *y*, with the increase being larger as *y* increases. Increasing *r* also increases average years of service, which approximately equals $1/(1 - r)$. If a service and Congress act to change *r*, what rate should be chosen and how does a single NCP affect this choice relative to a service-specific NCP?

The first-order condition for an efficient retention rate is

$$F_{1r_1} - [W_{1r_1}(1+\theta_1) + W_1\theta_{1r_1}] = 0,$$

where the subscript r_1 denotes the partial derivative with respect to the retention rate for service one. Force size and experience mix increase, which increases the pay bill and retirement liability, and increases capability. (Below, we allow *r* to increase but hold force size constant.) The first-order condition implicitly indicates the optimal size and experience mix. The full marginal cost is $[W_{1r_1}(1 + \theta_1) + W_1\theta_{1r_1}]$ while the marginal accrual cost under a service-specific accrual charge is $\theta_1 W_{1r_1} + W_1\theta_{1r_1}$. Recalling that

$$\theta_1 = \frac{L_1}{W_1},$$

we find

$$\theta_{1r_1} = \frac{L_{1r_1} - \theta W_{1r_1}}{W_1}.$$

Substituting this into the expression for full marginal cost simplifies it to $W_{1r_1} + L_{1r_1}$.

The first-order condition under a single NCP θ is

$$F_{1r_1} - [W_{1r_1}(1+\theta) + W_1\theta_{r_1}] = 0.$$

Here, the full marginal cost is $[W_{1r_1}(1 + \theta) + W_1\theta_{r_1}]$ while the marginal accrual cost is $\theta W_{1r_1} + W_1\theta_{r_1}$.

To compare marginal costs, we begin with the derivative of the single NCP with respect to r_1. With some math, it equals

$$\theta_{r_1} = \frac{L_{1r_1} - \theta W_{1r_1}}{W_1 + W_2}.$$

Substituting this into the marginal cost under the single NCP and simplifying, we have

$$(W_{1r_1} + L_{1r_1}) + (1-\alpha)(\theta W_{1r_1} - L_{1r_1}).$$

The term in the left parentheses is equal to the full marginal cost under the service-specific NCP. Therefore, the terms in the right parentheses give the bias from using a single NCP rather than the service-specific NCP. This can be of either sign, depending on whether θW_{1r_1} is greater or less than L_{1r_1}; both terms are positive.

Increasing Experience Mix Holding Force Size Constant

Assuming a steady state, force size equals the size of the entering cohort n times the sum of the fractions of the cohort retained to each year of service, which also equals the expected years of service over a 30–year career for an entering cohort of size n, computed earlier:

$$s = n\sum_{y=1}^{30} r^{y-1}.$$

We use this relationship to show the amount by which the size of the entering cohort must decrease to hold force size constant when the retention rate increases. In the preceding section, we saw that an increase in r would increase force size. It is useful to consider a change to experience mix (average years of service) holding force size constant. To find the tradeoff between n and r that holds force size constant, we totally differentiate the above expression, set it equal to zero, and solve for dn/dr:

$$ds = (\sum_{y=1}^{30} r^{y-1})dn + n[\sum_{y=1}^{30}(y-1)r^{y-2}]dr = 0$$

$$\frac{dn}{dr} = -\frac{n\sum_{y=1}^{30}(y-1)r^{y-2}}{\sum_{y=1}^{30} r^{y-1}}.$$

The numerator is the change in expected years of service from an incremental increase in r given an entry cohort of size n, and the denominator is the change in expected years of service from an incremental increase in n (adding a person to the entry cohort) given a retention rate r. The size of the tradeoff depends on n, which could be on the order of 50 or 100,000 service members, and on the retention rate, which is in the range of 0.9. Because of the huge scale difference between n and r, we express the tradeoff as the percent change in n required to offset a 1–percent increase in r:

$$\frac{dn / n}{dr / r} = -\frac{\sum_{y=1}^{30}(y-1)r^{y-1}}{\sum_{y=1}^{30} r^{y-1}}.$$

Approximating the right-hand side by assuming an infinitely long career, the numerator and denominator evaluate to $r/(1 - r)^2$ and $1/(1 - r)$, so their ratio is $r/(1 - r)$. For example, if $r = .9$ the trade-off is $.9/(1 - .9) = 9$, so the force size increase from a 1–percent increase in r is neutralized by a 9–percent decrease in entry cohort size.

The above discussion implies that under a single NCP a policy change that increases r but holds force size constant affects the change in accrual cost in two ways. By itself, the increase in r causes an increase in experience mix and force size, and this ex-post force size must be decreased to hold force size constant at its ex-ante level. The marginal accrual cost may be biased in either direction.

Change in Total Cost from Increasing Basic Pay for a Given Service

The change in total cost from an across-the-board increase in basic pay for service one under a service-specific and a single NCP (see the appendix) is:

$$\frac{dW_1(1+\theta_1)}{db} = W_{1b} + L_{1b} + (W_{1r} + L_{1r})\frac{\partial r}{\partial b}$$

$$\frac{dW_1(1+\theta)}{db} = \frac{dW_1(1+\theta_1)}{db} + (1-\alpha)[(\theta W_{1b} - L_{1b}) + (\theta W_{1r} - L_{1r})\frac{\partial r}{\partial b}].$$

The increase in basic pay b here can be thought of as a vertical shift in basic pay by year of service, e.g., an across-the-board increase of an *absolute* amount, and this increases r and the basic pay bill W_1.[8] The second equation implies that the change in total cost under a single NCP equals the change under a service-specific NCP plus two terms. Those terms can be of either sign, so the direction of bias is theoretically indeterminate. The example in Chapter Four presents a numerical estimate of the bias from the change in a service's pay relative to external pay; however, the example differs from the presentation here. There are two key differences: The derivation above uses an across-the-board change in basic pay of an incremental, absolute amount, while the example uses a percentage change in external pay; also, the derivation above is in the context of the heuristic model, while the example is based on the dynamic retention model and treats individual retention decisions as forward looking and strategic.

As an additional point, the change in accrual cost alone from an increase in basic pay under a single NCP has exactly the same bias as shown above (also, see the appendix).

How Would Marginal Cost Be Affected If Congress Always Covered Retirement Liability?

From above, the marginal cost with respect to r for a service-specific NCP is $n_1(W_{1r_1} + L_{1r_1})$. If Congress always covered a cohort's retirement liability, it would also be willing to fund an increased liability resulting from a higher retention rate. Because the service knows this, the retirement cost or saving from a change in its retention rate is not material to its choice of rate under a service-specific NCP. The term L_{1r_1} disappears, so in effect the service-specific NCP is zero and marginal cost is simply $n_1 W_{1r_1}$. Thus, marginal cost is lower at any given retention rate than it would have been with L_{1r_1} included. From a service's perspective, the lower marginal cost means that a higher retention rate

8 $W_{1b} = n_1 \sum_{1}^{30} r^{y-1} \beta^{y-1}$, and similarly θ_{1b},is the partial derivative of θ_1 with respect to b.

is optimal—assuming the accrual cost made a difference in the first place. A service's willingness to increase its force size and experience mix would only be limited by whether Congress would approve the increase in a basic pay bill. The same point applies under a single NCP if, as assumed, Congress is willing to cover any accrual charge.

It is simplistic to assume that Congress would cover the retirement liability regardless of its size, however. Otherwise, it would not have passed PL 98–94, instituting an accrual charge to provide current visibility for future retirement liability. Even though retirement benefit outlays occur in the future, Congress wanted to recognize the cost of providing for the benefits—if for no other reason than the resources have alternative uses either in the federal government or, with lower taxes, in the economy. Therefore, although Congress might ultimately cover an increase in liability, it would recognize that the underlying driver was the higher retention rate and would consider the benefit and cost of a larger, more senior force before deciding to cover the higher liability.

CHAPTER THREE

Insights from Past Studies

Introduction

This chapter draws from past studies' insights and recommendations about the current approach for funding military retirement costs. The studies we describe critique the single NCP and its application. The major shortcomings of the current approach that these studies highlight fall under the headings of inefficiency arising from inaccurate cost signals; weak incentives for efficient resource allocation; and unfairness shown by overestimating the retirement liability in the early years of the NCP system and, therefore, requiring accrual charges that were too high, but not returning funds to the services when this proved to be the case.

As background, under PL 98–94 any military retirement obligation incurred after October 1, 1984 is DoD's responsibility, and any obligation before that date is Treasury's responsibility. Each year, DoD contributes to the Military Retirement Fund (MRF), which Treasury maintains. The MRF is the financial vehicle for funding the retirement liability of those who have served. This transaction between DoD and Treasury is an intra-governmental transfer and does not represent an outlay (expenditure) of resources. That is, no federal spending occurs in the current fiscal year as a result of DoD's contribution to the MRF. Outlays only occur when Treasury makes payments to military retirees or their dependents. The amount DoD transfers to MRF each year is the accrual charge (also called normal cost), computed using the aggregate entry-age normal method.

The Office of the Actuary in DoD uses its "GORGO" model to compute the NCP. The computation requires a number of inputs and assumptions, summarized in the actuary's annual report *Valuation of the Military Retirement System*. The inputs and assumptions include estimates of future retention, and therefore the number of future active-duty members and military retirees, projections of pay growth, price growth for cost-of-living-adjustments, estimates of mortality rates, and an assumption about the rate of return on "funds" in the MRF.

In addition to its responsibility for military retirement obligations for service before October 1984, Treasury makes up any MRF deficits if DoD retirement liabilities were underestimated and receives any savings if DoD liabilities were overstated. These actuarial gains or losses to the MRF are amortized over a 30-year period. Differences between the projected liability and the actual liability may occur for a number of reasons discussed later in the chapter.

Gotz

Gotz (1985, unpublished) criticized the entry-age normal approach in his paper, *Military Retirement System Costing and Budgeting*. Key to his analysis was the insight that retirement accrual charges are the result of an accounting method and, importantly, not "synonymous with annually accruing retirement costs."

The NCP is based on a forcewide cumulative retention profile representing average retention across the services, but retention differs by service. The Marine Corps has the most junior force and the lowest percentage of personnel completing 20 years of service and qualifying for military retirement benefits, and the Air Force has the most senior force.

The use of a single NCP means that the accrual charge for a service with a relatively junior (senior) force, compared with the service-wide retention profile, would be higher (lower) than its actual accruing retirement liability. In other words, the use of a single, common NCP does not allocate accrual charges in relation to a service's accruing retirement liability. Furthermore, although service-driven changes in personnel and compensation policy can affect a service's retention profile and thereby affect its accruing liability, the single NCP mutes this effect. The service-driven changes can lead to change in the NCP, but because of averaging and a lagged adjustment in changing the representative profile, the change in the NCP will be less than if the NCP were service-specific and adjusted promptly. As a result, a service's accrual charge will not reflect the full cost increase or decrease of its personnel actions. The other military services will bear most of the increase or decrease; an increase or decrease in the single NCP affects them all.[1]

Gotz recommended adopting a system based on accrual cost (and not on an accrual charge based on an accounting method like entry-age normal) to address these shortcomings of the NCP accrual charge. In current-year dollars, the accrual cost of retirement equals the difference between a service's retirement liabilities at the end of year, which Gotz expressed as $(1 + r)PV_1(t)$,[2] and those at the beginning of the year, $PV_0(t)$. Liabilities at the end of the year would include the effect of changes in the service's personnel and compensation policy, as well as changes in actuarial assumptions (e.g., cost-of-living adjustments to retirement benefits, growth in basic pay, discount rate, and longevity) and experience gains or losses, which reflect the difference between the realized valued of the retirement fund and the a priori expectation. The accrual cost would be the budget charge:

$$AccrualCost = (1+r)PV_1(t) - PV_0(t).$$

Computation of the accrual cost would be service-specific. This approach would eliminate the disproportionate accrual charges that occur under the single NCP approach. Gotz also argued that by making the service's budget charge reflect the full increase or decrease in its liability, the service would have well-aligned incentives for an effi-

[1] In 1987, Congress mandated the transition to a separate NCP for part-time reservists (Hix and Taylor, 1997).

[2] This notation assumes that the effect of policy, assumption, and experience changes are realized at the beginning of the year in the term $PV_1(t)$, and the time-related cost of this amount for one year is the foregone interest, $rPV_1(t)$. Hence, the full cost is $(1 + r)PV_1(t)$.

cient experience mix of personnel. This approach, then, would address a major fault with the NCP accrual charge system: "...the DoD budget cannot correctly reflect the changes in accruing retirement liabilities resulting from changes in policy" (Gotz, unpublished).

Gotz does not discuss the relationship between the accrual cost in his approach and the objective of ensuring that the retirement liability is fully funded. The aggregate entry-age normal approach centers on the objective of fully funding the retirement liability, but as discussed it distorts information about a service's total and marginal cost of personnel and therefore provides inaccurate signals for resource allocation. Gotz's critique of the entry-age normal accrual charge is consistent with the idea that it does not consider the totality of the retirement liability, including past over- or underfunding relative to today's estimate of the liability. He places focus on the change in the liability, whereas the entry-age normal system focuses on ongoing charges to fund the liability subject to a representative retention profile.

One approach to considering the relationship between the Gotz approach and the entry-age normal approach is to recognize that, when a new cohort enters a service, the present value of the service's retirement liability would increase by the amount of the cohort's expected retirement liability. Therefore, this increase is included in the budget charge and might be its major component in many years; that is, the service would be charged for the full liability up front, at the time the cohort enters. The other portion of the accruing liability would come from experience gains or losses and assumption changes. If Gotz's method was in place on October 1, 1984 and if legislation had specified that DoD be responsible for retirement obligations arising from the military service of those *entering* from that day forward, then the accrual cost in each year from then on would include the liability of each entering cohort plus the liability change from changes in policy, assumptions, and experience.

Congress mandated that DoD have responsibility for the retirement liability for all service members, not just the liability for new entrants, starting October 1, 1984. Under these circumstances, Gotz's approach would have been consistent with a massive initial accrual cost for the liability for the entering cohort and DoD's share—somehow defined—

of the liability of each incumbent cohort. Alternatively, legislation could have given DoD credit in the MRF for its share of the liability of then-serving members, but this would be equivalent to charging DoD only for the liability of the entering cohort. Given that Congress made DoD responsible for retirement liability starting October 1, 1984, the entry-age normal system provided a way of allocating liability to Treasury and DoD. With a representative retention profile and the associated NCP, DoD would pay an annual charge equal to the NCP times its basic pay bill, and this would be expected to cover the post–October 1, 1984 accruing liability of incumbent personnel, with Treasury covering the liability for service up to that date. DoD's charge would be smaller under this approach because the NCP in effect amortizes liability over a 30–year career, given the retention profile. The entry-age normal approach had the effect of putting DoD retirement liability funding on a flow basis (an annual accrual charge) rather than a stock basis (paying the full expected liability at the time of a cohort's entry).

The flow-versus-stock comparison of entry-age normal accounting versus Gotz's approach can be seen in the relationship $L = \theta W$ from Chapter Two. This states that a flow—namely, the normal cost percentage times basic pay at each year of service for a cohort—if invested, is sufficient to cover a stock, the cohort's retirement liability. Gotz's approach would enter the cohort's liability into the annual accrual cost, while the entry-age normal system takes a share θ of basic pay at each year of service of a cohort to compute an annual accrual charge. From this perspective, Gotz's model and the entry-age normal approach share a common core—both can be expected to fully fund a cohort's retirement liability.

But beyond this, the approaches are starkly different. Gotz would have the change in a service's liability from one year to the next reflected in the accrual cost, whereas the entry-age normal approach would not. Specifically, the entry-age normal approach would be deficient in terms of the Gotz approach because the NCP is not service-specific and not adjusted promptly. Therefore, to the extent the NCP is adjusted to reflect gains or losses in the retirement liability, their full impact is not realized by the service in the current period but in effect is paid over time via the NCP.

Gotz's proposal to impose the full change in the present value of the retirement liability could create undesirable fluctuations in the budget charge. This could occur if the year-to-year change had a noticeable transitory component, i.e., random positive and negative changes that were largely independent of the "permanent" liability. The NCP, being more slowly adjusted, would dampen such fluctuations and make the budget charge more stable.

In Gotz's formulation, the accruing liability would include the effect of changes in assumption and experience gains and losses, but they may well be beyond a service's control. Including them would add non-Defense-related noise to the signal about the retirement liability. Assigning them to Treasury, as done under the current system, could avoid this. In that case, a service's accruing liability would be the result of accessing a new cohort of officers and enlisted personnel plus changes related to the retention of incumbent personnel. A challenge would be to determine whether such changes in a year were permanent or random and transitory. Again, the latter would be noise but could be hard to differentiate from permanent changes. Also, a service might want to argue that increases in its retirement liability were random and should not be counted, as a way of freeing funds for other uses in its budget. Claims of this sort would have to be evaluated within DoD and by Congress.[3]

If Gotz's proposal were modified to make Treasury responsible for liability changes resulting from assumption changes and experience gains and losses, and if the liability computation used an average of past continuation rates, then the accruing liability would be close to an accrual charge based on a service-specific entry-age normal system. That is, changing from today's aggregate entry-age normal to a service-specific entry-age normal would produce an accrual charge quite close to Gotz's accruing liability. Gotz wanted a service to be

[3] The DoD Actuary uses a servicewide retention profile based on year-to-year continuation rates averaged over several years. This retention profile adjusts slowly and is not responsive to a service's current actions to manage retention. However, the DoD Board of Actuaries could, if it chose, alter the NCP in response to structural changes, e.g., a change in the retirement benefit system.

fully and immediately responsible for the effect of its policy actions on its retirement liability. A single NCP cannot accomplish this.

Hix and Taylor

In *A Policymaker's Guide to Accrual Funding of Military Retirement*, Hix and Taylor (1997) argued for service-specific accrual charges. According to their assessment, a single NCP meant that the Army cross-subsidized the Air Force. They further argued for DoD and Treasury to share the gains and losses of the retirement fund.

As mentioned earlier, PL 98–94 assigned responsibility for retirement liabilities from military service before October 1, 1984 to the Treasury, and responsibility for liabilities on or after that date to DoD. It also assigned responsibility for retirement fund gains and losses to Treasury.[4]

As it happened, the retirement fund had actuarial gains of $288 billion in its first ten years. Gains of $166 billion came from experience, $117 billion from assumption changes, and $5 billion from benefit changes. Treasury amortizes gains and losses over 30 years,[5] but DoD did not share in those benefits because gains and losses were, by law, solely Treasury's responsibility. However, Hix and Taylor felt DoD should share in the gains and losses: "The law says that the monthly accrual payments are intended 'to permit the military services to recognize the full cost of manpower decisions made in the current year.'" Just as responsibility for retirement liabilities was split according to date, with DoD having responsibility for service on or after October 1, 1984, DoD should have responsibility for gains and losses occurring after that date, they argued.

---

[4] Actuarial gains and losses are estimated by computing fund assets and accrued liability at the end of a year, taking into account experience, assumption changes, and benefit changes, versus the beginning of a year.

[5] The retirement liability in 1985 was more than $500 billion. Treasury's responsibility was to pay off the liability in 60 years, i.e., the liability was amortized over 60 years. The amortizations of gains and losses over 30 years are adjustments to the 60-year amortization schedule.

They cited a 1992 report of the DoD Board of Actuaries recommending a change in the law to give DoD this responsibility, although no action was taken on this recommendation. This was reminiscent of a recommendation made when the accrual funding legislation was under discussion for the Department of Defense Authorization Act of 1984. The Report of the Committee on Armed Services (1983) called for the establishment of an annual accrual charge "levied against the Department of Defense budget as a whole or against the budget of each individual service." The report, in fact, recommended an individual service charge, arguing that this would allow the services "to recognize the full cost of manpower decisions."[6] But the law that was passed called for a "single percentage of basic pay," as we noted earlier.

Hix and Taylor gave two examples where Congress allowed DoD to share in cost savings from a decrease in accrual charges: the passage of REDUX in 1986 and the introduction of a separate NCP for part-time reservists in 1987. Both decreased the retirement liability and the retirement accrual charge. The accrual charge decreased because the NCP decreased. This meant that the stream of accrual charges in the current and future years would be lower; the NCP decrease was forward-looking.

The decrease in the retirement liability was a separate element and should be distinguished from the unfunded retirement liability. Accrual charges are used to fund the liability, and the unfunded liability is the difference between the current estimate of the liability and value of past accrual charges plus interest plus the present value of expected future accrual charges. Accrual charges for FYs 1985 and 1986 now were seen as too high because they were based on an initial estimate of the liability that was too high. But the extent of overcontribution was small because the accrual charge system had been in operation for only two years. The gain to DoD came not from a rebate on the 1985 and 1986 accrual charges but from lower future accrual charges. Importantly, DoD was allowed to use the savings from lower accrual charges "directly to fund [high-priority budget] items." That is, in this case the decrease in the accrual cost, an intra-governmental transfer,

---

6 We thank Saul Pleeter for bringing this to our attention.

was re-allocable to items in the budget requiring an outlay. The items were funded in order of their priority and "without regard to the service from which the savings had come." That is, gains from one service might have been spent on another service's procurement or operations.

In the second example, the expected decrease in the retirement liability from the 1990s drawdown was used to offset the costs of funding the Voluntary Separation Incentive (VSI) program, an incentive paid to facilitate the drawdown.

Hix and Taylor devote much of their report to the question of how to allocate changes in the unfunded retirement liability to Treasury and DoD, while Gotz built his on what he termed the accrual cost. This distinction exists because Gotz chose to focus on cost directly—his accruing liability is the change in the liability from one year to the next—whereas Hix and Taylor's change in unfunded liability is an actuarial estimate given the use of aggregate entry-age normal accounting. As we suggested in our comments on Gotz, this distinction, though conceptually clear, might not amount to much in practice if both approaches are service-specific, constrained to include gains or losses related to service decisions rather than to events outside service control, and based on the same retention profiles and economic assumptions. Both Gotz's and Hix-Taylor's approaches wanted service-specific assessments. Still, the distinction exists. Using our example of a new entry cohort (and no other changes, e.g., no experience gains or losses), Gotz's accruing liability increases because taking in the cohort implies an obligation to pay the cohort's expected future retirement benefits. But the service would be charged for this liability in its personnel budget, hence the accruing liability would be immediately funded. In the Hix-Taylor world, taking in the cohort not only increases the liability but also, through the entry-age normal system, simultaneously ensures that the liability is funded via the stream of accrual charges over the cohort's military career (as discussed below). So, other things equal, there should be no change in the unfunded liability.

The example of an entering cohort is useful because it illustrates that under the Gotz or Hix-Taylor frameworks, the cohort's retirement liability would be fully funded. But both frameworks also consider changes in the liability coming from changes in economic assumptions, experience, or service retention decisions. Hix and Taylor, in particular, seek to rationalize why a service should be able to retrieve past overfunding and their framework describes how this might be done. Gotz, in contrast, would have the service be fully responsible for its liability. The service would have to pay a cost in the current fiscal year equal to its accruing liability, which would be higher or lower if assumption changes or experience outcomes were worse or better than expected. If a service trimmed mid-career personnel from a cohort, its liability would decrease and this would be immediately reflected with a lower accruing liability. In contrast, in the Hix-Taylor context of entry-age normal accounting, the impact of the trimmed cohort would be felt only gradually as it was averaged into recent years' continuation rates.

Because gains and losses related to military service before October 1, 1984 were Treasury's responsibility and gains and losses after that date were DoD's responsibility, Hix and Taylor partition the change in the unfunded liability into changes from economic assumptions, experience, and benefits, with a separate calculation for each. The change in the unfunded liability for a given fund year (*t*) is

$$ChangeUFL_t = UFL_t(new) - UFL_t(old),$$

where *new* accounts for possible changes in actuarial assumptions (e.g., lifespan), experience (realization different from prior expectation), and benefit schedules, and *old* reflects the ex-ante counterparts. The unfunded liability equals the present value of future benefits ($PVFB_t$) for those now in the system, both retired and active, less the stream of future normal cost payments ($PVFNC_t$), less fund assets (F_t):

$$UFL_t = PVFB_t - PVFNC_t - F_t.$$

Putting the equations together,

$$\begin{aligned} ChangeUFL_t = {} & PVFB_t(new) - PVFB_t(old) + PVFNC_t(new) \\ & - PVFNC_t(old) + F_t(new) - F_t(old). \end{aligned}$$

The final step is to decompose each of these changes into changes for military service before versus after October 1, 1984. Hix and Taylor apply their framework in a logical way. For instance, an assumption change does not affect the value of the fund but can change the present values of future benefits and future contributions, so the fund terms drop out when computing the change in the unfunded liability in t caused from a change in assumptions. Similarly, normal cost charges took effect only for service on or after October 1, 1984, so all changes in $PVFNC_t$ are assigned to DoD.

The Hix-Taylor decomposition takes the entry-age normal accrual system as a given. This can be seen in the terms for the present value of the new and old normal cost streams. Also, by including the change in the value of the fund (new fund value minus old fund value), they bring fund gains and losses into the analysis. The change in the present value of benefits also enters the picture.

In the entry-age normal system, the DoD Board of Actuaries periodically evaluates the retirement liability and the adequacy of the NCP for funding it, and adjusts the NCP if needed. Under PL 98–94, "Not less often than every four years, the Secretary of Defense shall carry out an actuarial valuation of Department of Defense military retirement and survivor benefit programs," which shall include "a determination (using the aggregate entry-age normal cost method) of a single level percentage of basic pay for active duty" and "a determination (using the aggregate entry-age normal cost method) of a single level percentage of basic pay and of compensation ... for members of the Selected Reserve of the armed forces." Chapter Two derived the formula for the NCP, $\theta = L/W$, from which it follows that changes in L or W alter the NCP. The liability can change if benefits, interest rate, longevity, inflation, basic pay, or the percentage of the cohort reaching 20 years of service shift. The basic pay bill can change if there are revisions to basic pay, interest rate, inflation, or retention. The updated NCP is applied going forward; there are no retrospective adjustments to past accrual charges, and the law offers no provision for a return of funds to DoD or a service and, in fact, makes Treasury responsible for actuarial gains and losses to the military retirement fund—despite also making DoD responsible for retirement liabilities for service as of October 1, 1984, as Hix and Taylor point out.

Table 3.1 shows the Hix-Taylor allocation of changes in the unfunded retirement liability. We derived the table from several tables in their report. The overarching result is that DoD should be responsible for changes in the unfunded retirement liability for all service from October 1, 1984 onward. This includes changes in the value of future benefits, as well as changes in the value of future normal-cost accrual charges, regardless of whether the changes stem from changes in assumptions, experience, or benefits.

The Hix-Taylor approach raises questions of implementation. One issue in allocating gains and losses to Treasury versus DoD is that some changes affecting the unfunded liability may be outside of a service's control. The same point applies to Gotz's approach. Should a service be held responsible for changes in the unfunded liability caused by factors beyond its decision authority, such as changes in assumptions about inflation, interest rate, and longevity? Similarly, is a service responsible for changes in civilian wage opportunities and basic pay that change retention, which in turn changes retirement liability? Military services have virtually no control over civilian wages, and the services and Congress act together when proposing and enacting military compensation changes. Furthermore, Congress, not the services, ultimately decides on budget authorization and appropriation. Congress has not assigned a property right or obligational authority to services that would allow them to retrieve and spend actuarial gains to the retirement fund (Hix and Taylor) or a positive difference between expected and realized accruing liability (Gotz). Similarly, actuarial losses or a negative difference between expected and realized accruing liability cannot credibly be assigned to a service when it is Congress that authorizes and appropriates funds and has the power to increase or decrease a service's budget. These arguments undercut the case that DoD or a service owns the accrual cost funds transferred to the military retirement fund and can draw upon them when there is a surplus.

But in the Hix-Taylor framework, the service would retrieve the large decrease in the estimated retirement liability in the initial years after 1984. Suppose that an actuarial valuation in 1988 discovered the

Table 3.1
Hix-Taylor Allocation of Changes in the Unfunded Retirement Liability

Component	Assumption[1]		Experience[2]		Benefit	
	DoD	Treasury	DoD	Treasury	DoD	Treasury
Pre-84 PVFB		X		X		X
Post-84 PVFB	X		X		X	
PVFNC	X		X		X	
Fund	NA		Apportioned by past fund contributions		Zero value	

NOTES: PVFB = present value of future benefits; PVFNC = present value of future normal cost payments.

[1] Changes in assumptions concern the assumed values of basic pay growth, cost-of-living adjustments, and interest rate.

[2] Experience gains or losses also depend on basic pay growth, cost-of-living adjustments, and interest rate but reflect realizations versus expected outcomes given the economic assumptions.

overestimate and revised down the NCP. As mentioned, these actuarial gains were mainly from assumption changes and experience.[7] The services, having paid accrual charges in 1984 through 1987 that turned out to be higher than needed, would "own" the decrease in the liability for military service after October 1, 1984. Yet they could not spend this gain without congressional approval. One possibility is that Congress would preapprove spending the surplus. However, this seems unlikely because Congress might want to consider the causes and circumstances at the time the surplus is realized; and, even with preapproval, Congress might change its mind and override the action. In addition, if a

[7] It might seem appropriate to allocate gains and losses to a service when they are based on authority delegated to the service, such as for force structure and experience mix. But even decisions about force size and experience involve interactions external to the service: service proposals are reviewed in DoD's planning and budgeting process and are subject to congressional approval. Also, although a service's actions can change its unfunded liability, other contemporaneous changes not under a service's control can also change it. If so, there may be no clear way to determine what part of the change in the unfunded liability should be allocated to the service.

"deficit" appeared (the NCP that had been charged was too low), it does not make sense for the service to own the deficit because it is Congress that decides on the service budget.

A more-subtle interpretation of the Hix-Taylor framework is that it identifies gains or losses that could be allocated to a military service, while the actual allocation depends on the interaction between the service and Congress. This is consistent with Congress taking into consideration past accrual charges when it passed legislation to fund the drawdown incentives VSI and Special Separation Bonus and when it allowed DoD to increase its spending on high-priority items once the initial overfunding in 1985 and 1986 was discovered. These cases also fit with another point raised by Hix and Taylor, namely, that a decrease in another part of a service's budget should not negate its gain from a decreased retirement liability. "If DoD is permitted to begin sharing in gains, it obviously stands to benefit the most if its aggregate budget level does not decline correspondingly and it is instead allowed to spend the difference on other priorities."

Finally, the decision to declare a gain or loss requires judgment about whether it is transitory or permanent. For example, gains from changes in the interest rate or inflation might be temporary. Similarly, slower growth in basic pay decreases the retirement liability, but pay must eventually be set at a competitive level if manning requirements are to be met. Like Gotz, Hix and Taylor did not discuss how to identify transitory versus permanent changes. In the current system, the DoD Board of Actuaries monitors the retirement fund and its assumptions and makes adjustments periodically, which guards against overresponse to temporary fluctuations.

Eisenman, Grissmer, Hosek, and Taylor

Eisenman and his colleagues (2001) document research done before Hix and Taylor (1997). The research provides a rationale for applying the decrease in the retirement liability from the drawdown toward funding the VSI and critiques the aggregate entry-age normal accrual method as it was implemented, arguing that several factors weakened its capacity to serve as an incentive for efficient personnel management. The NCP was based on forcewide retention and masked differences in the retirement liability accrued by different service and communities. Conservative economic assumptions and a failure to account for the drawdown led to high estimates of the retirement liability and unnecessarily high accrual charges during the first decade of the accrual method (1985–1994). All gains and losses to the unfunded liability were assigned to Treasury, preventing the military services from sharing in the gains materializing during the first decade. Finally, the adjustments to the NCP that did occur were hard to predict and, because gains and losses were amortized, muted the impact of current changes in personnel policy affecting the retirement liability.

The report recommended making accrual charges service-specific, separate for officers and enlisted personnel, and specific to each entering cohort.[8] Accrual charges should be promptly updated in response to changes in the retirement liability[9] and take into account not only past retention outcomes but also predicted future retention, which could adjust with policy revisions. The services should be responsible for changes in their retirement liability that "might properly be attributed" to them, including gains and losses in their unfunded liability. In particular, gains and losses should be assigned to specific services rather

---

[8] There are already separate normal cost percentages and accrual charges for AC and RC forces, in particular, for "regular" and "part-time" personnel.

[9] Eisenman and his colleagues (2001, p. 24) wrote,

> The accrual payment is affected directly and immediately by changes in payroll caused by shifts in experience mix and the level of compensation, but the NCP is almost immune to annual changes in personnel plans. Thus, planners generally regard the NCP as exogenous to the personnel planning process. For instance, if personnel planners were to tighten permanently pre-retirement tenure rules to restrict the number reaching retirement and thereby raise accession levels to keep force size constant, lower present and future payrolls would result and should trigger a lower NCP. Under present methods used by the actuaries, this policy action would lower the payroll in the following year, thereby lowering DoD contributions. However, it would not be reflected in payroll projections for future years nor would it affect the current NCP until years later. The reason is that the projections of future-force structure and the cohort calculations leading to the NCP use continuation rates from 5–15 years ago. Thus, any current change in continuation rates would not be fully reflected for 15 years.

than to DoD. Finally, to make the accrual charge into a meaningful incentive, the authors recommend advance funding of the retirement liability and permitting the funds to be expendable by the service.

This research shares the limitations of Hix and Taylor that we have discussed. A distinctive feature is the recommendation for advance funding of the retirement liability, which parallels Gotz given our interpretation of how his approach might be implemented. The authors observe that the fungibility of retirement funds has been "governed in each instance through negotiations between DoD, Office of Management and Budget, and Congress. Thus, decisionmakers cannot plan for use of savings—but neither are they required to reduce funding [applied to other purposes in their budget] in cases of accrual increases." Hence, "the only way that fungibility can be guaranteed to the services is to convert the retirement fund into an advance-funding accrual method. Establishing a trust fund with real dollars would remove any distinction between accrual expenditures and all other expenditures." Yet, as mentioned, Congress is unlikely to make this precommitment.

Eisenman and his colleagues also recommended the use of a side account to return gains and losses to the services on an amortized basis. This is the same approach Treasury uses to handle gains and losses. The side account can be seen as a way to avoid rewriting the 1984 legislation establishing the use of entry-age normal accounting. But the NCP already adjusts to changes in the present value of the retirement liability and basic pay bill, so the existing system seems to largely meet the purpose of the side account.

Dahlman

Dahlman's (2007) *The Cost of a Military Person-Year: A Method for Computing Savings from Force Reductions* distinguishes between the accrual charge for funding a retirement liability and the economic cost of an additional year of service from an entry cohort of military personnel. Dahlman's focus is within the service career of a cohort. This contrasts to Gotz, Hix and Taylor, Eisenman and colleagues, and the model in Chapter Two.

Dahlman argued that the entry-age normal method is conceptually inappropriate for determining the cost of another year of service from a cohort and therefore is a misleading guide for service decisions regarding experience mix. He would instead assign retirement cost "to each [year of service] based on the retention patterns induced by the cliff-vesting system; in other words, the costs of accrual should be distributed in proportion to the probability of each year-group actually reaching retirement eligibility. Using this methodology, the accrual costs of year-groups with small retirement probabilities will be assigned a lower cost than year-groups with high retirement probabilities." For service after 20 years, when everyone is eligible for retirement benefits, an additional liability comes from benefit increases resulting from more years of service and promotions. Dahlman would allocate the post-20 accrual cost according to the expected retirement liability given the year of service at retirement, after allowing for the cost of the "basic retirement package," which covers the cohort's retirement liability up to 20 years of service.

For instance, if $\bar{b}_{20}$ is an entering cohort's expected average basic pay at 20 years of service, the cohort's expected retirement liability for the basic package is approximately

$$L = n_{20} \times .025 \times 20 \times \bar{b}_{20} / \delta \ ,$$

where n_{20} is the number of year-group members expected to reach 20 years of service.[10] The amount L would be apportioned by year of service in accord with the probability of reaching 20 years of service given the year of service. If $s_{20|j}$ is the probability of reaching 20 years of service from year of service j, the accrual cost assigned to that year is

[10] This is the amount at year 20, and it can be discounted back to earlier years of service. Discounting and the return on investment are omitted for simplicity. Assuming the government discount rate and rate of return on funds are the same, the math in the text is accurate.

$$\frac{s_{20|j} L}{\sum_{y=0}^{19} s_{20|y}}.$$

The accrual cost for post–20 service is determined as in the following example. If k_{21} is the number of cohort members retiring at 21 years of service, their retirement charge equals their retirement liability in excess of the basic package:

$$k_{21} \times .025 \times (21 \times \bar{b}_{21} - 20 \times \bar{b}_{20}) / \delta.$$

This is done for all post–20 retirement years, and the sum of retirees over those years equals the number eligible to retire, n_{20} (ignoring the possibility of death).

Dahlman argued that his method provides an accurate signal of the economic cost of each year of service over an entering cohort's life-cycle of service. Intuitively, the closer a member is to retirement eligibility at 20 years of service, the more likely the retirement liability will be realized. By allocating the accrual charge in proportion to the probability of reaching 20 years of service, the method tells decision-makers that personnel costs increase nonlinearly according to year of service up to 20 years. Dahlman also recognized that at the extreme of this logic, the retirement liability would be realized only at exactly 20 years of service—this, because of cliff vesting at 20. But rather than have an abrupt increase in cost at 20 years and nothing before, he believed it was reasonable to spread the cost increase over all 20 years in proportion to the chance of reaching 20 years conditional on current year of service. In contrast, the level percentage of the aggregate entry-age normal accounting applies the same percentage of basic pay at each year of service. Also, it applies the same NCP to service after as well as before 20 years of service, which means it does not recognize the change in accruing liability that occurs once eligibility has been attained.

It is interesting to contrast Dahlman's method to Gotz's method. In Gotz's method (as we interpret it), the retirement liability of an incoming cohort of personnel would be in the personnel budget at the time of accession. Barring assumption changes and experience gains and losses, this means no further accruing liability would arise during the service career of the cohort—zero cost for accruing liability after entry. In Dahlman's method, which in effect is an alternative accounting method to entry-age normal, the annual charge by year of service is in proportion to the probability of reaching 20 years of service from that year.

If in the current fiscal year a service decided to separate a member at year of service 15, Gotz's accruing liability would decrease by the present discounted value (as of the current fiscal year) of the member's expected retirement benefits, a calculation taking into account the probability that the member would have stayed in service to 20 years and perhaps longer. That is, Gotz's approach fully recognizes the decrease in liability and registers it immediately in the current fiscal year, the year when the member is separated.

Using the notation for Dahlman above, this amount would be $S_{20|15}L$ for a member who would have stayed 20 years and then left. Dahlman would show a current-year decrease in the accrual cost equal to

$$(s_{20|15} / \sum_{y=0}^{19} s_{20|y})L,$$

a considerably smaller amount. Dahlman and Gotz would register these decreases in the current fiscal year only. To extend the comparison to entry-age normal, the accrual cost would decrease by $NCP \times b_{15}$ in the current year, $r_{16} \times NCP \times b_{16}$ the next year, $r_{16}r_{17} \times NCP \times b_{17}$ the year after that, and so on. The accrual cost savings would be the expected entry-age normal accrual cost avoided. Dahlman's approach would allow a downsizing military service to reap a greater accrual cost decrease than would the entry-age normal approach. Yet both Dahlman's approach and the entry-age normal approach are accounting methods, and the change in accrual cost under these methods is not the change in the actual cost. Only Gotz's method shows that. Gotz's method provide an accurate signal of the change in cost from downsizing, which is an important datum for policy decisions, whereas the other methods show the change in accrual cost given their respective accounting conventions.

The different methods might have different appeal from the point of view of budgeting. Suppose a service wanted to argue that retirement cost avoided should be reprogrammable into other parts of the budget, e.g., for operations. Gotz's method—by capturing the present discounted value of the decrease in retirement liability—would allow the service the largest possible increase in its operations budget and result in federal outlays increasing by that amount. The allowable increase would be much lower under Dahlman's approach and still lower under entry-age normal.

Dahlman compared his approach to Standard Military Composite Rates (SMCRs), which are grade-specific cost factors for military personnel developed by the DoD Comptroller. "Under DoD regulations, these are the rates to be used for use in cost calculations undertaken for civilianization purposes." SMCRs increase according to pay grade, but not as fast up to 20 years of service as the conditional probability of reaching 20 years. Also, after 20 years of service, the SMCR for officers is higher than Dahlman's cost, and is at first higher and then lower for enlisted members than Dahlman's cost. The rates for the SMCR use the current entry-age normal approach to compute retirement costs. Thus, the comparisons with Dahlman's costs are not surprising.

From Dahlman's perspective, (a) the entry-age normal approach assigns too low of a cost as years of service near 20 and too high of a cost after 20 years, and (b) SMCRs assign too low of a cost as years of service approach 20 and has a mixed relationship with Dahlman's cost after 20 years. Because both approaches increasingly underestimate cost as years of service approach 20, Dahlman argued that force planners do not have the information they need to allocate manpower efficiently. Because pre-20 senior personnel appear cheaper than they are, the force will be too senior up to 20; and because post-20 personnel appear more expensive than they are, there will be too few.

By assigning a higher accrual cost to personnel approaching 20 years of service, Dahlman's method helps to highlight the added cost, or cost savings, of changing the experience mix of an incumbent cohort and specifically the percentage to retain to 20 years of service. Decreasing this percentage decreases the cohort's retirement liability, and Dahlman's method would bring a larger and more immediate decrease in the accrual charge than would entry-age normal. This leads to the question: Would the service be able to spend these cost savings, i.e., would the accrual cost savings be converted into a spendable amount? The answer depends on whether the service's top-line budget guidance would be held constant (at one extreme) or adjusted downward by the amount of the accrual charge decrease (at the other). As before, the outcome depends on the interaction between the military service and Congress, the service's proposed use of the funds, and competing uses for funds.

Hogan and Horne

Hogan and Horne's (1989) *The Military Retirement Accrual Charge as a Signal for Defense Resource Allocation* focuses on the effect of entry-age normal accrual accounting on resource allocation and economic efficiency. At the time they wrote, accrual charges had been too high, leading to an inflated cost of Army manpower. This was not their main concern, however, as increasing the Army budget to compensate could correct it. Instead, they worried that accrual accounting gave the wrong signal to decisionmakers with respect to the tradeoff between labor and capital within and between the services. Decisionmakers faced the wrong marginal costs.[11]

---

[11] Hogan and Horne (1989), on p. 6, wrote,

> Two consequences follow from inflated manpower costs in the Army. First, as manpower is perceived to be more expensive relative to equipment, more equipment will be substituted for manpower.... Secondly, Army manpower will appear more expensive relative to the Navy and Air Force.

They illustrate how a procurement decision could be affected by the single NCP approach. The labor input from an Army E–3 soldier appears to be more expensive than it really is, which could lead the Army to economize on the use of E–3 soldiers and purchase automatic loaders when, in fact, costs would be less if the Army expanded tank crews to include an E–3 soldier to load rounds. When they wrote, the cost of E–3 personnel over the life cycle of the loader was actually less, not more, than the cost of the loader. Although a broader analysis would include the training and coordination cost of manning a four- versus three-member tank crew as well what resources were needed to train the additional personnel and how they would be used after they are E–3s, nevertheless their example is illustrative of the potential misallocation under a single NCP.

Hogan and Horne's discussion of the marginal cost of manpower focuses on the budget cost of a person-year. But other marginal costs could be relevant depending on the specifics of the question at hand. For instance, there is the marginal cost of an accession holding expected lifecycle retention constant, the marginal cost of increasing lifecycle retention in some way, and the marginal cost of higher retention to a particular year of service but not after (as induced by a separation bonus), and so forth. The model in Chapter Two illustrates several marginal costs, for instance; these are the marginal cost of an increase in retention (without an increase in military pay), the same but holding force size constant, and the marginal cost of an increase in pay.

Conclusion

Overall, we derive several insights from this review. First, in defense of the aggregate entry-age normal system, bringing visibility to the retirement liability associated with current manpower decisions did improve on the PAYGO approach. This was a chief objective. Second, however, it has shortcomings. The accrual charge is not accurate at the service level. All services bear the increase or decrease in the retirement liability resulting from one service's decisions. The NCP is adjusted periodically and, as a result, accrual charges are not immediately responsive to changes in the liability. (An exception is when the retirement system itself changes, in which case a new NCP is computed for service under the new system.) The services cannot lay claim to actuarial gains to the retirement fund, e.g., from past overfunding.

All of the studies stressed the importance of clear cost signals for efficient resource allocation, and in particular Gotz, Hix and Taylor, and Eisenman and colleagues recommended service-specific costs or accrual charges. This would address the critiques that the NCP was inaccurate at the service level and that all services bore the increases or decreases in the liability caused by one service. Moreover, both the total and the marginal accrual costs would be appropriate to the service. As a result, it could eliminate grumbling and suspicion among the branches that service A pays for service B's retirement liability and, as a result, service A's budget is being squeezed, leaving less room for spending elsewhere.

Gotz's method addresses the critique that accrual charges are not responsive to changes in liability. By putting the accruing liability in the service's budget rather than an accrual charge based on the aggregate entry-age normal system, Gotz's method would make the full change in the liability immediately part of the budget. A downside of this method is that random swings in the liability would be felt in the budget, causing variation irrelevant to the "true" liability. The entry-age normal approach avoids this; the NCP is updated periodically, by law at least every four years. Also, as we argued, in a steady policy environment a service-specific NCP-based accrual charge would be close to the accruing liability Gotz proposed, though without the randomness.

But the periodic adjustment and steadiness of the NCP have the downside of delaying the accrual charge adjustment to current changes in policy that increase or decrease a service's liability and, when the NCP has been adjusted, stretching out the impact of the change in liability over time though higher or lower future accrual charges. We offer a response to this: The policy analysis supporting a service's decision that will affect its retirement liability should include an estimate of the change in liability. This is typically done in practice and provides decisionmakers with the relevant cost signal.

Gotz, Hix and Taylor, and Eisenman and his colleagues argued that the services should be responsible for their accruing retirement liability and, by that reasoning, should be able to claim past overfunding for current spending. Hix and Taylor proposed a framework for allocating actuarial gains and losses to DoD or Treasury, but they did not present a fiscal framework to accomplish this. Eisenman and colleagues proposed creating a trust fund that a service could draw on if there were actuarial gains. But if there were losses, the same reasoning calls for a "fund" to pay off the losses. We have argued that approaches allowing the services to share in the gains and losses of the military retirement fund are not implementable without a commitment from Congress assigning the service a property right to the gains and losses. But this is not credible because a service on its own has no source of revenue to pay for the losses (the service depends on Congress for its funding), and because Congress cannot bind itself (a future Congress can reverse a past Congress' commitment). In our view, the services and Congress must work together as they do anyway in the annual budget cycle.

Finally, Dahlman criticized the entry-age normal method because it did not accurately portray the increase in expected retirement liability by year of service, which increases nonlinearly toward 20 years of service. Dahlman's critique falls under the heading of inaccurate cost signals. Dahlman's method would increase the accrual charge toward 20 years of service. Dahlman's method also would provide a stronger signal than entry-age normal to force planners of the expected retirement liability savings or cost from decreasing or increasing retention at, say, year 16 versus year six. However, it would make no difference in assessing the ex ante retirement liability of an entering cohort, as that depends on the percentage of personnel expected to reach 20 years of service. Moreover, Dahlman's accounting cost would not be as accurate a signal as Gotz's accrual liability. As suggested above, any proposed policy action from the services on retirement liability should include an estimate of the change in the liability. This would be a clear, direct cost signal for policy and would avoid confusing the accrual charge with the cost of the action. Still, both Dahlman's and Gotz's methods would apparently allow a service today to capture more of the expected future cost savings of trimming senior personnel from the force. Hix and Taylor, and Eisenman and his colleagues, because they work within an entry-age normal framework, would have such savings accrue gradually in the form of lower annual accrual costs. But under any of these schemes, whether a service can obtain cost savings from a decreased retirement liability depends on Congress. Congress could decrease the service's budget by the amount of the savings, or alternatively, allow the service to keep all or part of the savings. The point again is that the service and Congress interact on determining resource amount and allocation.

CHAPTER FOUR

Estimates Using Service-Specific NCPs Versus a Single NCP

The Office of the DoD Actuary computes the NCP used for the retirement accrual and would be called on to compute service-specific NCPs. However, we also provide estimates of the NCP for each service with the use of a model RAND researchers have developed, the DRM, which permits simulations of new and untried policies to see the effect on AC retention, RC participation, and cost for both active and reserve components. We have described the model in past studies and only briefly describe it here before turning to the results of the analysis.

As discussed in Chapters Two and Three, a single NCP does not accurately allocate accrual charges in relation to each service's retirement liability. Service-driven changes in personnel and compensation policy can affect retention and a service's retirement liability, but a single NCP mutes this effect.[1] This is because the NCP is based on a representative retention profile that, in effect, averages over the retention profiles of all services and the adjustment of the NCP is lagged and based on historical retention rates. Therefore, a service would not capture all of the savings from decreasing its liability, or bear all of the costs from increasing its liability, and the savings it does capture would

[1] In the previous chapter, we said that an estimate of a change in the liability should accompany policy decisions affecting the retirement liability, thereby providing a direct signal of the liability-related cost of the decision. This chapter focuses on accrual charges. Although accrual charges are not economic costs, they are the "costs" appearing in personnel budgets and are real in that sense, and we estimate the extent to which a single NCP results in inaccurate total and marginal accrual costs.

occur over time through decreased accrual charges. As a result, the current system does not provide clear signals and incentives for an efficient mix of experienced personnel.

The estimates in this chapter show a service's accrual charge under a single NCP and a service-specific NCP, one for enlisted and one for officers. The difference in these estimates measures the extent to which the use of a single NCP causes a service to be under- or overcharged for its accrual. There are also estimates of marginal accrual cost under single and service-specific NCPs. The empirical results are consistent with the model in Chapter Two and discussion in Chapter Three: a single NCP biases a service's total and marginal accrual cost.

Overview of Our Approach

RAND's DRM is well suited to the analysis of structural changes in military compensation. Recent applications include analyses of the ninth, tenth, and 11th Quadrennial Reviews of Military Compensation (QRMCs) and analyses of retirement reforms under consideration by DoD and the one recommended by the MCRMC (Mattock, Asch, and Hosek, 2014). The model's capability has steadily increased. For instance, new, faster estimation and simulation programs have been written, cost has been refined, and the model can now show retention and cost effects in both the steady state and the year-by-year transition to the steady state.

The model is based on a mathematical model of individual decisionmaking over the life cycle in a world with uncertainty, and its parameters are estimated with data on military careers drawn from administrative data files. The main version of the model begins with service in the AC, and individuals make a stay-leave decision in each year. Those who leave the AC take a civilian job and, at the same time, choose whether to participate in the RC. Each year, a reservist can choose to remain in the RC or to leave it to be a civilian, and a civilian can choose to enter the RC or remain a civilian.[2]

[2] For a more complete discussion of the model, see Asch, Hosek, and Mattock (2013).

Our retention data are from the Defense Manpower Data Center Work Experience File (WEX). The WEX contains person-specific longitudinal records of active and reserve service. We used WEX data for service members who began their military service in 1990 or 1991 and tracked their individual careers in the AC and, if they joined, the RC through 2010, providing 21 years of data on 1990 entrants and 20 years on 1991 entrants. For each AC component, we drew samples of 25,000 individuals who entered the component in FYs 1990 and 1991, constructed each service member's history of AC and RC participation, and used these records in estimating the model. We supplemented these data with information on active, reserve, and civilian pay. AC pay, RC pay, and civilian pay are averages based on the individual's years of AC, RC, and total experience, respectively. We used 2007 military pay tables, and we do not expect our results to be sensitive to the choice of year because military pay tables have been fairly stable over time.[3] For civilian pay opportunities for enlisted personnel, we used the 2007 median wage for full-time male workers with associate degrees. For officers, we used the 2007 80th percentile wage for full-time male workers with master's degrees in management occupations. The data on civilian pay opportunities are from the U.S. Census Bureau.

In the model, each service has a single RC. The Army National Guard and U.S. Army Reserve are not treated separately but are combined into a single group, the Army RC, and similarly for the Air National Guard and Air Force Reserve. The model assumes that military pay, promotion policy, and civilian pay are time stationary, and it excludes demographics such as gender, marriage, and spousal employment, as well as health status, health care benefits, deployment, and deployment-related pays. That said, the estimated models fit the observed data extremely well for the both the AC and the RC.

We estimated the model separately for officers and enlisted in each of the four military services. We also developed simulation code

---

[3] An exception was the structural adjustment to the basic pay table in FY 2000 that gave larger increases to mid-career personnel who had reached their pay grades relatively quickly. A second exception was the expansion of the basic allowance for housing, which increased in real value from FY 2000 to FY 2005.

allowing us to simulate retention over the military career in both the AC and RC and compute the cumulative retention profile in the steady state. We simulate the retention profile under the current compensation system, what we call the baseline force, and we can also simulate retention under hypothetical alternative compensation systems, such as reforms to the retirement system or changes to basic pay.

A feature of our simulation capability is the computation of personnel costs. Our cost estimates include the cost of current compensation, including basic pay, the housing allowance and the subsistence allowance, and the cost of retired pay for the steady state, baseline force.

We use the cost element of our simulation capability to estimate by military service, for officers and enlisted personnel, the active and reserve NCP and accrual costs under the single NCP and, hypothetically, under service-specific NCPs.

Total Accrual Costs Under Single and Service-Specific NCPs

Table 4.1 shows the AC service-specific NCPs and accrual costs for officers and enlisted personnel in each of the four military services. As mentioned, the cost estimates are based on the simulation results for the steady state under the current retirement system. We caution that the estimates in Table 4.1, based on our model, are not official estimates. That is, the DoD Office of the Actuary makes official estimates of the accrual charge for the military retirement system, but the actuary does not currently compute a service-specific or officer and enlisted NCP within each service. Our estimates from the DRM are an approximation of what the DoD Actuary would estimate for officers and enlisted within each service. To benchmark our estimates to the Actuary's NCP under the current system, we scale our estimates, and, consequently, the difference in accrual charges between our estimates and the actuary's NCP is zero under the current system, by construction. Overall, we believe our estimates are accurate approximations. Also, our focus in this analysis is the NCP faced by each service in DoD, whether a single NCP or a service-specific NCP. The DoD Actu-

Table 4.1
AC NCPs and Costs Under Service-Specific NCPs for Enlisted and Officers Versus Single NCP (2015 $billions)

Service	Rank	DoD 2015 NCP	Estimated NCP	NCP Delta	PB FY15 Basic Pay	DoD Accrual Charge	Service Accrual Charge	Difference in Accrual Charge
Army	Enlisted	32.2%	24.5%	7.7%	$15.06	$4.85	$3.69	$1.16
Army	Officer	32.2%	41.7%	(9.5%)	$7.90	$2.55	$3.29	$(0.74)
Army Total					$22.96	$7.40	$6.98	$0.42
Air Force	Enlisted	32.2%	38.8%	(6.6%)	$8.94	$3.47	$2.88	$(0.59)
Air Force	Officer	32.2%	44.2%	(12.0%)	$2.44	$1.08	$0.78	$(0.29)
Air Force Pilots	Officer	32.2%	42.8%	(10.6%)	$2.44	$1.04	$0.78	$(0.26)
Air Force Total					$13.82	$5.59	$4.44	$(1.14)
Navy	Enlisted	32.2%	27.2%	5.0%	$8.92	$2.43	$2.87	$0.44
Navy	Officer	32.2%	38.5%	(6.3%)	$4.14	$1.59	$1.33	$(0.26)
Navy Total					$13.06	$4.02	$4.20	$0.18
Marines	Enlisted	32.2%	17.0%	15.2%	$4.92	$0.84	$1.58	$0.75
Marines	Officer	32.2%	45.1%	(12.9%)	$1.54	$0.69	$0.50	$(0.20)
Marines Total					$6.46	$1.53	$2.08	$0.55

SOURCE: Authors' calculations.

NOTES: Totals are rounded; parentheses used for negative totals. PB = President's Budget ary computes an NCP for DoD, equal to 32.2 percent in 2015, as well as an NCP for DoD plus Treasury, equal to about 42 percent. The difference between the two is the cost of concurrent receipt, a cost that Treasury bears rather than DoD.[4]

For the Army, we estimate a service-specific NCP of 24.5 percent for enlisted personnel and 41.7 percent for officers. As mentioned, DoD's single NCP in 2015 is 32.2 percent. Thus, the service-specific NCP for enlisted personnel is 7.7 percent points lower, while that for officers is 9.5 percentage points higher. The lower NCP for enlisted personnel and higher NCP for officers result from Army enlisted retention being below, and Army officer retention being above, retention in the representative retention profile across the services. Figure 4.1 illustrates Army officer and enlisted retention profiles relative to the common retention profile. As seen, Army enlisted retention to 20 years of service is below the common retention profile, while Army officer retention is above it.

A key insight from our model, as well as from past studies, is that the use of a single NCP will overstate a service's manpower budget cost if its service-specific NCP is lower than the single NCP, as is the case with Army enlisted personnel. Similarly, it will understate a service's manpower budget cost if its service-specific NCP is higher, as is the case with Army officer personnel. The question is, how much is the overstatement and understatement, and in the case of the Army, what is the net effect for both enlisted personnel and officers?

Table 4.1 provides some estimates to address these questions. We use basic pay bills of $15.06 billion and $7.9 billion for Army enlisted personnel and officers, respectively for full-time personnel in 2015 dollars, as provided by the Office of the Secretary of Defense Actuary and

---

[4] Prior to 2004, a member's military retirement benefits were offset (i.e., reduced) by any amount of VA disability benefits the retiree might receive, because members are not allowed to concurrently collect both types of compensation. This offset was eliminated in 2004 for those with a 50-percent or greater disability rating with the Concurrent Retirement and Disability Pay (CRDP) program. CRDP allows eligible military retirees to receive both military retired pay and VA compensation. The cost of CRDP is borne by Treasury, resulting in a total NCP that differs from the NCP for DoD, and faced by each military service.

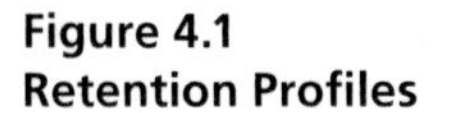

Figure 4.1
Retention Profiles

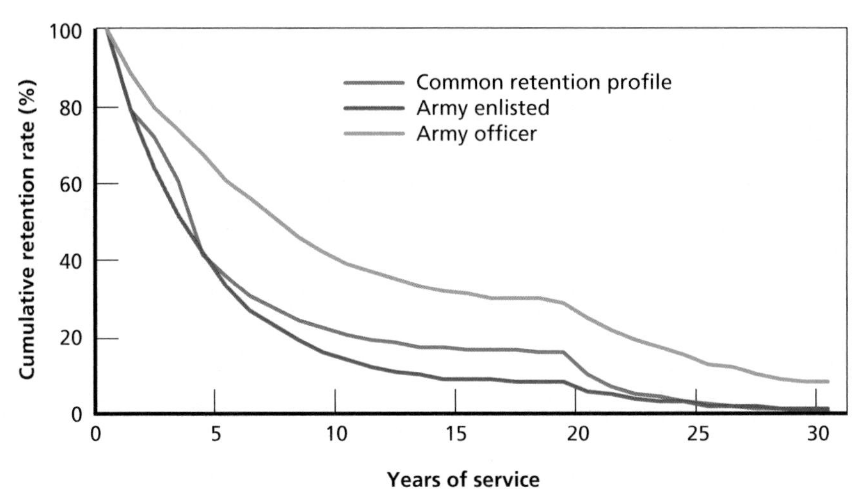

SOURCE: Authors' calculations.
RAND *RR1373-4.1*

Comptroller and included in the 2015 President's Budget. Applying the Army enlisted NCP of 24.5 percent produces an enlisted accrual charge of $3.69 billion, which compares with $4.85 billion under the single NCP. Thus, the Army's accrual charge for enlisted personnel would be $1.16 billion less under an Army-specific enlisted NCP. That is, the overstatement of retirement costs is $1.16 billion for full-time Army enlisted personnel. The numbers for Army officers show that the accrual charge would be $750 million higher under an Army-specific officer NCP. That is, the understatement of costs for Army officers is $750 million. For officers and enlisted personnel together, the Army's accrual charge would be $407 million less than under the single NCP. Thus, on net, Army retirement costs are overstated by $407 million for full-time personnel in using the single NCP rather than the service-specific NCPs for officers and enlisted personnel in the table. Although the NCP delta is higher for Army officers (–9.48 percent) compared with the NCP delta for enlisted personnel (7.68 percent), the bias for Army enlisted personnel outweighs the bias for officers because the enlisted basic pay bill is nearly double that for officers, reflecting the larger size of the Army enlisted force.

The Marine Corps would have nearly as large a decrease in accrual charges if service-specific NCPs for enlisted personnel and officers were used. Although the Marine Corps is smaller than the Army—its $6.46 billion basic pay bill is less than a third of the Army's $22.96 billion bill—the Marine Corps has the most junior experience mix, and its enlisted NCP is only 17.0 percent. Its officer retention profile is similar to that of the Air Force, but the Marine Corps has a lower officer-enlisted ratio. Therefore, the Marine Corps' accrual charge would be $551 million less under a service-specific NCP. The Navy's retention profile is closest to the representative retention profile, but it too would have a $183 million lower accrual charge under a service-specific NCP for officers and enlisted personnel.

Not surprisingly, the Air Force's accrual charge, in contrast, would increase. Both the enlisted and officer NCPs for the Air Force are greater than the single NCP. Thus, using a service-specific NCPs for officers and enlisted personnel would increase the Air Force enlisted retirement costs $590 million and its officer retirement costs $550 million.[5] On net, Air Force retirement costs would increase by more than a billion ($1.140 billion) if service-specific NCPs for officers and enlisted personnel were used. Or, put differently, the single NCP approach leads to an understatement of Air Force costs of about $1 billion.

The higher accrual cost for the Air Force leads to questions and concerns among Air Force planners about whether legislation that would move to a service-specific NCP would also include increased funding for the Air Force to cover this increased cost. The issue of how to transition to a service-specific NCP will be addressed in Chapter Five. We next present estimates of retirement marginal cost from changes in force size and experience mix under the single versus service-specific NCPs.

---

[5] As discussed in Asch, Hosek, and Mattock (2014), we estimate separate models for non-rated and rated Air Force officers. Thus, Table 4.1 shows separate results for these two subgroups of Air Force officers.

Estimated Marginal Accrual Cost with Respect to Force Size Under Single and Service-Specific NCPs: Current Retirement System

In this subsection, we first provide estimates of the marginal accrual cost from an increase in force size under a single versus service-specific NCP. We then consider the marginal accrual cost under a policy example, namely a force downsizing for the Army that decreases experience mix. As expected, the estimates show that the marginal accrual cost with respect to force size for enlisted personnel for the Army is overstated under the single NCP approach, so the incremental retirement cost of Army enlisted personnel appears too high. Furthermore, we find that downsizing results in a decreased incremental cost of Army enlisted personnel, but the decrease is understated under the single NCP approach. That is, the Army would not receive full credit for the decline in its cost from downsizing; the incremental decrease in cost is actually larger than what is estimated using the single NCP approach.

The computation of marginal cost in our analysis here is different and simpler than in the Chapter Two model. Discounting the expected pay bill and the expected retirement liability were critical to the model to determine the NCP; the NCP is the level percentage of basic pay such that, taken at each year of service and invested, the amount available will grow enough to cover the retirement liability. However, here we are interested in accrual cost in the current-year budget. In current-year dollars, the accrual charges are $W_1\theta_1$ and $W_1\theta$ in the service-specific NCP and single NCP approaches where W_1 is *not* discounted and is simply the basic pay bill for the current force.

That said, the formulas in Chapter Two guide the computation of marginal cost. The marginal accrual cost (for a force size change holding retention or experience constant) is $\theta_1 W_{1n_1}$ under a service-specific NCP and $\theta_1 W_{1n_1}$ under a single NCP approach. Given the lack of discounting, the term W_{1n_1} is given by

$$W_{1n_1} = \sum_{y=1}^{30} r^{y-1} b_y .$$

Thus, W_{1n_1} is simply the average basic pay for the current force, and our estimate of the marginal accrual cost is equal to average basic pay times the NCP.

Table 4.2 shows the estimated marginal accrual cost under the service-specific and common single NCP approach for an incremental increase in force size without a change in experience mix or retention. The single NCP leads to a biased marginal accrual cost. For the Army, it is biased up for enlisted personnel, $8,057 under an Army enlisted NCP versus $10,580 under the single NCP. Because of this, force planners have an incentive to reduce the enlisted force size. This is also true for the Marine Corps and Navy. The reverse is true for officers for all military services, because the officer NCP is higher than the single NCP. The single NCP underestimates the marginal accrual cost of officers, and force planners have an incentive to increase officer force size.

In the case of the Air Force, the estimated marginal accrual cost is too low for both enlisted personnel and officers. For example, the marginal cost is $13,178 under an Air Force enlisted NCP versus $10,934 under a single NCP. Thus, the Air Force has an incentive to increase the sizes of its enlisted and officer forces.

Estimated Marginal Accrual Cost Under Single and Service-Specific NCPs with Respect to Experience Mix: Current Retirement System

We consider the case of an Army downsizing that decreases experience mix. If downsizing were strictly proportional across years of service, the percentage of personnel remaining at each year of service would be unchanged and so would the NCP. The downsized force would be smaller but have the same experience mix. If the force were smaller by 10 percent, for example, the accrual charge would be 10 percent less (see Chapter Two) because the basic pay bill would be smaller. In this case, marginal accrual cost with respect to experience mix would be unchanged.

But if downsizing changed the experience mix the NCP would change, as would the marginal accrual cost. Our example illustrates

Table 4.2
Estimated Marginal Accrual Cost With Respect to Force Size Under Service-Specific NCPs for Enlisted Personnel and Officers and a Single NCP (2015 $)

Service	Enlisted/ Officer	Service-Specific NCP	Single NCP
Air Force	Enlisted	$13,178	$10,934
Army	Enlisted	$8,057	$10,580
Marines	Enlisted	$4,580	$8,691
Navy	Enlisted	$8,924	$10,557
Air Force	Officer	$32,889	$23,953
Air Force Pilots	Officer	$31,815	$23,953
Army	Officer	$36,284	$28,034
Marines	Officer	$33,484	$23,904
Navy	Officer	$30,635	$25,608

SOURCE: Authors' calculations.

this case. To simulate an Army downsizing that changes experience mix, we consider a 1-percent increase in the civilian wage for the Army only. In reality, a change in civilian pay would affect retention in all services, but our purpose here is not to show how changes in civilian pay affect military retention; we use the civilian pay increase only as a device to achieve Army downsizing without affecting other services.

Figure 4.2 shows the effect of a permanent, 1-percent civilian pay increase on Army enlisted retention in the steady state. The black line is the number of Army enlisted members serving on active duty by years in service in the steady state under the baseline scenario and the red line is the same when civilian pay is 1 percent greater. The simulation holds accessions fixed, so the impact of the civilian wage increase is to rotate the Army enlisted retention curve downward (decrease the curve more as years of service increase). This is like the model in Chapter Two where an increase in caused the retention profile to rotate upward, so

Figure 4.2
Army Downsizing Resulting from a 1-Percent Increase in the Civilian Wage

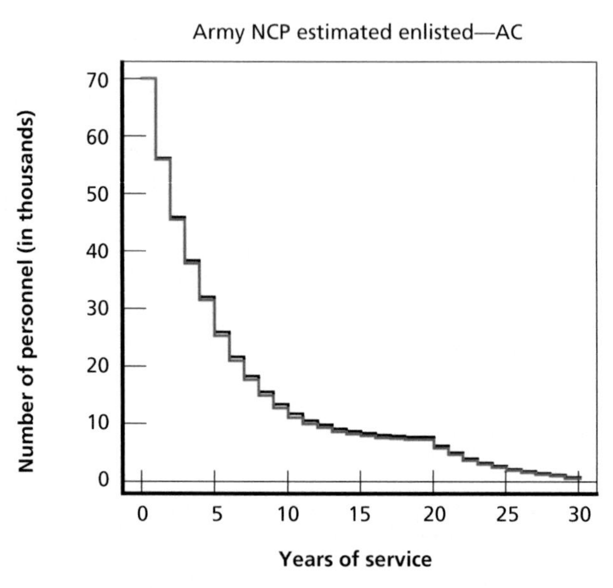

RAND *RR1373-4.2*

the findings from the DRM and the model should be similar (though in opposite directions).

The enlisted force size falls by about 18,000. In a similar simulation for Army officers (not shown), force size falls 1,800. As expected, the experience mix decreases. Man-years per accession fall 0.25 years for enlisted and 0.27 years for officers.

Table 4.3 expands on Table 4.2 to include columns for the marginal accrual cost with respect to force size evaluated at the ex-post retention curve, which reflects the decrease in experience mix. As in Table 4.2, the marginal accrual cost for Army enlisted is higher under a single NCP than under an Army-enlisted–specific NCP. Also, the marginal accrual costs for enlisted and officers with respect to force size follow the same pattern as in Table 4.2, as expected.

Furthermore, Table 4.3 has columns showing the difference in the marginal accrual cost before and after the decrease in experience mix. The table shows the effects on the marginal accrual cost when experience mix changes, but only for the Army. The marginal retirement cost differs when the experience mix changes both because average basic pay falls when the force becomes more junior and because the NCP decreases as fewer entrants reach retirement eligibility. This can be understood as follows. Recall that a service's total accrual cost in the model was θW, and the marginal accrual cost of increasing force size was the derivative of this with respect to n, or θW_n. (As above, however, W is not discounted.) Now the incremental change with respect to r:

$$\frac{\Delta(\theta W)_n}{\Delta r} = \theta \frac{\Delta W_n}{\Delta r} + W_n \frac{\Delta \theta}{\Delta r}.$$

Note that we can write W_n as

$$W_n = \frac{W}{n}.$$

So, the above expression becomes

$$\frac{\Delta(\theta W_n)}{\Delta r} = \frac{1}{n}\left(\theta \frac{\Delta W}{\Delta r} + W \frac{\Delta \theta}{\Delta r}\right).$$

In other words, the incremental change in the marginal accrual cost is simply $1/n$ times the marginal cost with respect to experience mix. Thus, the difference in the marginal accrual cost shown in the right-most columns of Table 4.3 is the per entrant marginal accrual cost with respect to a change in force mix.

With a service-specific NCP, this figure for Army enlisted is (\$7,388–\$8,057=) –\$669. Scaling this figure to the size of the force, the incremental change in retirement cost from downsizing is -\$294.7 million in 2015 dollars. Under the single NCP, the comparable figure is (\$10,251-\$10,580=) -\$329. Scaling this figure, the incremental change in retirement cost is smaller, about –\$144.9 million. For Army officers, the marginal cost with respect to a decrease in experience is (\$35,536-\$36,284=) –\$748. This scales to –\$66.6 million for the

Table 4.3
Estimated Marginal Accrual Cost Under a Single NCP and Service-Specific NCPs for Officers and Enlisted Personnel Before and After Army Downsizing (2015 $)

		Before Downsizing		After Downsizing		Difference	
Service	Enlisted/ Officer	Single NCP	Service-specific NCP	Single NCP	Service-specific NCP	Single NCP	Service-specific NCP
Air Force	Enlisted	$10,934	$13,178	$10,749	$13,178	($185)	$0
Army	Enlisted	$10,580	$8,057	$10,251	$7,388	($329)	($669)
Marines	Enlisted	$8,691	$4,580	$8,544	$4,580	($147)	$0
Navy	Enlisted	$10,557	$8,924	$10,378	$8,924	($179)	$0
Air Force	Officer	$23,953	$32,889	$23,548	$32,889	($405)	$0
Air Force Pilots	Officer	$23,953	$31,815	$23,548	$31,815	($405)	$0
Army	Officer	$28,034	$36,284	$27,406	$35,536	($627)	($748)
Marines	Officer	$23,904	$33,484	$23,499	$33,484	($405)	$0
Navy	Officer	$25,608	$30,635	$25,175	$30,635	($433)	$0

SOURCE: Authors' calculations.

NOTES: Parentheses used for negative amounts.

Army officer force. Under a single NCP, the marginal cost is –$627, which scales to about –$55.9 million. Thus, the actual decrease in the Army's retirement cost when it downsizes 20,000 is $361.3 million per year ($294.7 million + $66.6 million), but the single NCP approach makes it appear as though the Army cost decreases only $200.7 million per year. That is, the Army only receives a little more than half of the credit for its budgetary cost decrease.

Interestingly, not only do the Army's marginal accrual costs change when the Army downsizes, so do the marginal accrual costs of the other services under the single NCP approach. For two services, the single NCP (Chapter Two) is

$$\theta = \frac{W_1\theta_1 + W_2\theta_2}{W_1 + W_2} .$$

When the Army downsizes and becomes more junior, its own NCP, θ_1, and the average wage bill, W_{1n}, decrease. The effect on the numerator is larger than the denominator, so θ falls. In our example, the single NCP falls from 32.2 percent to 31.7 percent. As a result, the other services have a lower accrual charge as well. The DoD total accrual cost decrease in our example under the single NCP will appear to be $382.4 million, of which the Army gets credit for $200.7 million.

Clearly, the current NCP approach does not accurately reflect cost changes when the Army changes its force size and shape. While our examples are for the Army, the qualitative result would be the same for the other services if each alone downsized and decreased seniority. Consequently, all of the services face incorrect cost incentives when changing their experience mix.

Estimated Total and Marginal Accrual Costs Under Retirement Reform

Numerous past commissions and studies have reviewed and recommended reform of the military retirement system. Most recently, DoD convened a working group that reviewed the system and recommended two alternative approaches for reform in a white paper it released in 2014 (DoD Office of the Actuary, 2014), while the MCRMC presented its recommendations for reform in its final report issued in 2015 (MCRMC, 2015). These recommended reforms differ in their details, but they share a common structure. In each case, they involve reducing the current defined benefit pension and supplementing it with a defined contribution plan with government contributions that would vest earlier but pay out benefits later in a member's lifetime than the defined benefit plan and with higher current compensation. In the case of the MCRMC, the higher current compensation takes the form of the addition of continuation pay targeted to members in their mid-career. The defined contribution plan would specifically be the Thrift Savings Plan (TSP) that covers federal employees, though the specifics of that plan would not necessarily be identical to the plan offered to federal employees. Finally, reducing the retirement benefit multiplier accomplishes the decrease in the defined benefit pension. Under the current formula, the benefit an eligible member receives is given by 2.5 percent times years of service times the average of highest three years of basic pay. Both the MCRMC and one of the approaches proposed in the DoD review recommended reducing the multiplier from 2.5 percent to 2.0 percent.

Analyses of these recent proposals are presented in the DoD white paper and MCRMC final report as well as in supporting documentation (Asch, Hosek, and Mattock, 2014; Asch, Mattock, and Hosek 2015). But in summary, the key findings are that these proposals can sustain active and reserve component retention, reduce steady state costs, increase the percentage of military entrants who will become vested, and offer the opportunity for more flexible management of military personnel.

One implication of these reform proposals is that the accrual charge is reduced. Clearly, as can be seen in the heuristic model, reducing the retirement multiplier reduces the accrual charge. But, the accrual charge is reduced even if the defined contribution plan contributions are funded on an accrual basis and included in the accrual charge.[6] The result is that accrual costs are found to be lower under retirement reform, and indeed, personnel costs are lower for each service under retirement reform even after incorporating the higher cost associated with higher current compensation.

The question of interest from the perspective of the current analysis is whether the cost savings to the services of retirement reform would differ under a system of service-specific and enlisted/officer-

[6] The MCRMC envisioned that the defined contribution plan contributions would be paid directly by DoD to the TSP Retirement Board, rather than to the military retirement fund, so it would not be funded on an accrual basis. However, early costing of the MCRMC recommendation involved incorporating these contributions into the accrual charge, and costing of the DoD approaches in the white paper also involved incorporating TSP contributions into the accrual charge. In both cases, the accrual charge decreased.

specific accrual charges relative to the cost savings under the current single accrual charge system. A related question is how the disparities in the marginal cost signals change. Not surprisingly, marginal costs as well as total costs are lower under retirement reform. As shown above, the single NCP approach produces an inaccurate cost signal to the services when making force size and experience mix decisions. The question at hand is how retirement reform affects the size of this inaccuracy. We address these questions in this subsection.

The retirement reform we consider is the MCRMC proposal, though the results we present would generalize qualitatively to the DoD approaches outlined in the 2014 white paper (MCRMC, 2014) or other proposals with a similar structure to the MCRMC and DoD reforms that would reduce the retirement accrual charge while sustaining retention. Specifically, the proposal we consider here has the following general features for the active duty retirement system:

- Defined benefit retirement program, vested at 20 years of service with immediate benefits for vested members, using a formula of 2 percent × years of service × average of the highest three years of basic pay. Under the MCRMC proposal, members can choose to receive a lump sum in place of all or part of the annuity between the age at retirement and age 67; though, for simplicity, we assume in this analysis that all members receive an annuity between the age at retirement and age 67.
- Defined contribution retirement program, vested at the beginning of the third year of service, with an automatic DoD contribution of 1 percent of basic pay starting at the beginning of the first year of service and ending at the end of year 20 of service. The program also has a matching element, with DoD matching member contributions up to 5 percent of basic pay, starting at the beginning of the third year of service. For the purpose of the analysis here, we assume all members contribute 3 percent, thereby receiving a 3–percent match.
- Continuation pay paid to members at 12 years of service, as a multiplier of monthly basic pay, to sustain retention.

The cost in the MCRMC final report assumes a portion of the force would choose the lump sum option while the cost here assumes none would choose the lump sum. We make this assumption to simplify the analysis.[7] Thus, the costing here is not directly comparable to what is shown in the MCRMC report or in RAND's supporting documentation (Asch, Mattock, and Hosek, 2015).

We reproduce Table 4.1 to start the discussion, but instead assuming the MCRMC retirement system rather than the current system. The new table shows the single and service-specific NCPs under the MCRMC proposal and illustrates how total budgetary retirement costs differ under the retirement reform, just as they differ under the current system. Next, we show how the cost savings associated with retirement reform at the service level changes under a service-specific accrual charge. That is, the overall DoD-wide cost savings associated with retirement reform are unchanged, but the allocation of those savings across the services differs under a service-specific accrual system. Finally, we show marginal cost estimates under retirement reform and compare them with the estimates under the current system. Not surprisingly, marginal costs are lower, but the inaccuracy in these cost signals changes under retirement reform. A key finding is that the services will continue to face inaccurate cost signals with retirement reform under the single NCP approach, but the size of the inaccuracy is reduced.

Total Accrual Costs Under Single and Service-Specific NCPs Under Retirement Reform

Table 4.4 shows the AC service-specific NCPs and accrual costs for officers and enlisted personnel in each of the four military services under retirement reform. The Office of the Actuary estimates that the single NCP under the MCRMC proposal, assuming all members

[7] We make this assumption because the results we present will depend on how the lump sum is computed (e.g., which discount rate is used) and what percentage of the force will choose the lump sum. The MCRMC left open the issue of how the lump sum would be computed, leaving it to the DoD Actuary to define. Because of this uncertainty about the lump sum and the effect it has on the results, we show the simplest case where all are assumed to choose an annuity.

choose an annuity (and assuming a 3–percent TSP government matching rate) would be 27.4 percent. This is shown in the third column of the table. The table also shows the service-specific NCPs under retirement reform for officers and enlisted personnel. These estimates are based on DRM simulation results for the steady state. Comparing the NCPs with those in Table 4.1, both the single NCP and the service-specific NCPs are lower under retirement reform than under the current system, even when the NCP under retirement reform includes the TSP contributions.

For the Army, we estimate a service-specific NCP of 20.5 percent for enlisted personnel and 35.8 percent for officers. Thus, the service-specific NCP for enlisted personnel is 6.93 percentage points lower while that for officers is 8.38 percentage points higher. More generally, similar to the current system, the service-specific NCPs are lower than the single NCP for enlisted personnel, with the exception of the Air Force, and are higher than the single NCP for officers. Again, this result stems from the differences between the representative retention profiles across the services and the service-specific and enlisted- or officer-specific retention profiles. We also find that the overstatement and understatement in the budget figures found in Table 4.1 from using a single versus a service-specific NCP are slightly smaller in absolute value under retirement reform. For example, the overstatement in the Army's budget cost from using a single NCP is smaller, $381 million rather than $407 million. Similarly, the understatement in the Air Force budget is lower, $1.016 billion rather than $1.141 billion.

A question of particular interest to the services is what are the cost savings they will realize under retirement reform and how would those savings differ under a service-specific NCP approach. To answer that question we need to compare the accrual costs in Tables 4.1 and 4.4 and account for the change in personnel costs associated with sustaining retention under the retirement reform through the use of continuation pay. Table 4.5 shows the relevant cost comparisons. Specifically, it shows total accrual costs under the current system and total accrual costs plus the costs of continuation pay under the retirement reform, and these costs under the single NCP and service-specific NCP

Table 4.4
AC NCP and Steady State Costs Under Single NCP Versus Service-Specific NCPs for Enlisted and Officers Under Retirement Reform (2015 $billions)

Service	Enlisted/ Officer	Single NCP	Service-specific NCP	NCP Delta	PB FY15 Basic Pay	Single Accrual Charge	Service-specific Accrual Charge	Difference in Accrual Charge
Army	Enlisted	27.4%	20.5%	6.93%	$15.06	$4.13	$3.08	$1.05
Army	Officer	27.4%	35.8%	(8.38%)	$7.90	$2.17	$2.83	$(0.66)
Army Total								$0.39
Air Force	Enlisted	27.4%	33.3%	(5.95%)	$8.94	$2.45	$2.98	$(0.53)
Air Force	Officer	27.4%	38.0%	(10.56%)	$2.44	$0.67	$0.92	$(0.26)
Air Force Pilots	Officer	27.4%	36.7%	(9.29%)	$2.44	$0.67	$0.89	$(0.23)
Air Force Total								$(1.02)
Navy	Enlisted	27.4%	22.9%	4.54%	$8.92	$2.45	$2.04	$0.41
Navy	Officer	27.4%	32.9%	(5.47%)	$4.14	$1.13	$1.36	$(0.23)
Navy Total								$0.18
Marines	Enlisted	27.4%	14.6%	12.82%	$4.92	$1.35	$0.72	$0.63
Marines	Officer	27.4%	38.8%	(11.39%)	$1.54	$0.42	$0.60	$(0.18)
Marines Total								$0.45

SOURCE: Authors' calculations.

NOTES: Totals are rounded; parentheses used for negative amounts. PB = President's Budget.

approaches for enlisted personnel and officers. Finally, it shows the cost savings or the differences under each approach in the last two columns.

The last row of Table 4.5 shows that the DoD-wide cost savings associated with the retirement reform is $1.87 billion, and this amount is the same whether a single NCP or service-specific NCP is used to compute the savings. Thus, from the standpoint of DoD and Congress, the cost savings are the same—$1.87 billion under both approaches.

That said, the savings realized by each service for enlisted personnel and officers differs under the different NCP approaches. Under the current single NCP approach, the Army would realize a $640 million savings for enlisted personnel in the steady state compared with $520 million under an approach that uses an NCP specific for Army enlisted personnel. In contrast, for officers, it would realize $160 million under the single NCP but $240 million under the service-specific NCP. More generally, except for the Air Force, the single NCP approach leads to an overstatement of the change in total costs for enlisted personnel and an understatement of the change for officers. In the case of the Air Force, the single NCP leads to an understatement of the savings for enlisted personnel as well as officers. Again, this is because the retention profile for Air Force enlisted personnel lies above the representative one that is the basis for the single NCP. On net, more of the cost savings would go to the Air Force under the service-specific NCP approach than under the single NCP approach. In other words, the use of a single NCP means that the cost savings are understated for the Air Force and overstated for the other services.

Of course, the services will not realize any difference in savings between a single and service-specific NCP if Congress adjusts the service budgets to reflect the differences shown in Table 4.5. The more important issue from the standpoint of efficient resource allocation is whether the single NCP approach will continue to reflect inaccurate cost changes when the military services change their force sizes and shape under retirement reform as they do under the current system.

Estimated Marginal Accrual Cost with Respect to Force Size Under Single and Service-Specific NCPs: Retirement Reform

Table 4.6 compares the estimated marginal accrual cost for an incremental increase in force size under the current retirement system (replicating the figures in Table 4.2) with retirement reform. The marginal cost estimates under retirement reform are uniformly lower than under the current system. Thus, retirement reform lowers both total and mar-

Table 4.5
Steady State Cost Savings Under Retirement Reform Under Single NCP Versus Service-Specific NCPs for Enlisted and Officers (2015 $billions)

		Retirement Reform		Current		Difference	
Service	Enlisted/ Officer	Service-specific Accrual Cost + Continuation Pay Cost	Single Accrual Cost + Continuation Pay Cost	Service-specific Accrual Cost	Single Accrual Cost	Service-specific	Single
Army	Enlisted	$3.17	$4.21	$3.69	$4.85	$(0.52)	$(0.64)
Army	Officer	$3.05	$2.39	$3.29	$2.55	$(0.24)	$(0.16)
Army Total						$(0.77)	$(0.79)
Air Force	Enlisted	$3.04	$2.50	$3.47	$2.88	$(0.43)	$(0.37)
Air Force	Officer	$1.01	$0.75	$1.08	$0.78	$(0.07)	$(0.03)
Air Force Pilots	Officer	$0.98	$0.75	$1.04	$0.78	$(0.06)	$(0.03)
Air Force Total						$(0.56)	$(0.44)
Navy	Enlisted	$2.12	$2.53	$2.43	$2.87	$(0.31)	$(0.35)
Navy	Officer	$1.50	$1.28	$1.59	$1.33	$(0.09)	$(0.06)
Navy Total						$(0.40)	$(0.40)
Marines	Enlisted	$0.74	$1.37	$0.84	$1.58	$(0.09)	$(0.21)
Marines	Officer	$0.64	$0.47	$0.69	$0.50	$(0.05)	$(0.03)
Marines Total						$(0.14)	$(0.24)
Grand Total		$18.13	$18.13	$16.26	$18.13	$(1.87)	$(1.87)

NOTE: Parentheses used for negative amounts.

ginal accrual costs, implying that personnel are uniformly less expensive in terms of retirement costs at the margins under the retirement reform. Still, as under the current system, the single NCP leads to inaccurate marginal accrual cost under retirement reform. For the Army, it is too high for enlisted personnel: $6,726 for an Army enlisted NCP versus $9,003 for the single NCP under reform. The final two columns show the difference in marginal accrual costs between the service-specific NCP approach versus the single NCP approach under the current system and under retirement reform. Under the current system, the difference for Army enlisted is $2,523 (=$10,580–$8,057). Under retirement reform, the difference is $2,277 (=$9,003–$6,726). Thus, under retirement reform, the inaccuracy, while still present, is reduced. The implication is that while the force planners still have an inefficient incentive to reduce the size of the enlisted force (since the marginal cost appears too large under the single NCP approach), that incentive is a bit weaker under the retirement reform. We find a similar result for enlisted personnel in the Navy and Marine Corps.

As under the current system, the NCP is too low for officers for all services and for enlisted personnel in the Air Force under retirement reform. For example, for Army officers, marginal accrual cost is $31,152 under an Army officer NCP versus $23,855 under a single NCP. As before, this occurs because the officer NCP and the Air Force enlisted NCP is higher than the single NCP, even under retirement reform, though those differences are smaller, as shown in Table 4.6. The single NCP underestimates the marginal accrual cost of officers, and the services have an incentive to increase officer force size. The difference between the single NCP and service-specific NCP is smaller under retirement reform. In the case of Army officers, the difference is –$7,297 ($23,855–$31,152) rather than –$8,250 under the current system.

Table 4.6
Estimated Marginal Accrual Cost Under a Single NCP and a Service-Specific NCP Under Current Retirement System and Retirement Reform, for Officers and Enlisted Personnel

		Retirement Reform		Current Retirement System		Difference	
Service	Enlisted/ Officer	Single NCP	Service-specific NCP	Single NCP	Service-specific NCP	Retirement Reform	Current System
Air Force	Enlisted	$9,304	$11,323	$10,934	$13,178	($2,019)	($2,244)
Army	Enlisted	$9,003	$6,726	$10,580	$8,057	$2,277	$2,523
Marines	Enlisted	$7,395	$3,935	$8,691	$4,580	$3,460	$4,111
Navy	Enlisted	$8,983	$7,494	$10,557	$8,924	$1,489	$1,633
Air Force	Officer	$20,382	$28,240	$23,953	$32,889	($7,858)	($8,936)
Air Force Pilots	Officer	$20,382	$27,296	$23,953	$31,815	($6,914)	($7,862)
Army	Officer	$23,855	$31,152	$28,034	$36,284	($7,297)	($8,250)
Marines	Officer	$20,340	$28,796	$23,904	$33,484	($8,456)	($9,580)
Navy	Officer	$21,791	$26,142	$25,608	$30,635	($4,352)	($5,027)

SOURCE: Authors' calculations.

NOTES: Parentheses used for negative totals.

CHAPTER FIVE
Conclusion

Accrual costing of the military retirement liability has been in use for 30 years. It has brought visibility to the accruing liability through accrual charges. Despite the growth in the retirement liability under the large standing military in the Cold War era, the previous system, based on PAYGO funding for retirement benefits, delinked force size and structure decisions from their implications for retirement liability. The introduction of accrual charges changed this in creating a direct link. A higher liability now results in higher accrual charges in the current personnel budget.

PL 98–94 called for aggregate entry-age normal accounting and required a single-level percentage to be applied to basic pay in computing the accrual charge. This mandate clearly did not intend for the method to be service-specific, but rather was meant to reflect the retirement liability of all the military services together. But this led to high accrual charges for the Army, Navy, and Marine Corps, and a low accrual charge for the Air Force, relative to what those charges would be under service-specific accruals. In addition, as we have shown, when policy actions change a service's size or experience mix, the incremental changes in the service's accrual charge are different from what they would be under service-specific accounting. Simply put, the current system provides inaccurate estimates of the total and marginal accrual charges at the service level. The inaccurate signals are manifested in budget items—accrual charges—that are too high or too low and therefore are not as helpful as they could be for efficient resource allocation.

Technically, with present-day data systems, software, and computing power, fewer obstacles stand in the way of shifting to service-specific accrual charges. Data exist to construct retention profiles for enlisted and officers separately by service, and this is at the heart of implementing service-specific accruals. Accounting elements including the return on funds, inflation rate, basic pay growth, benefit changes, and longevity are the same across services.[1] Furthermore, the initial objectives of accrual counting legislation would still be met. Service-specific accrual charges would bring visibility to the retirement liability and ensure funding.

Past studies have discussed additional issues with the accrual accounting system as established under PL 98–94. Basically, these argue for allowing a service to retrieve past overfunding to spend in its current budget, receive credit today for manpower management decisions that decrease its retirement liability, obtain these gains sooner via faster adjustment of the NCP, and the use of an alternative accounting approach that assigns increasing high accrual charges as years of service approach 20. While sympathetic to the rationale for these suggestions, we have raised questions about their implementation. Because the military services and Congress reach consensus on budgets jointly (with Congress as the ultimate fiscal authority), a service cannot be said to own its past overfunding or solely responsible for any underfunding. Also, Congress cannot constrain itself, and while it can credibly commit to pay future retirement benefits as mandatory spending, Congress has little reason to commit to including actuarial gains or losses in the military retirement fund in the current budget, which is discretionary spending. Quicker adjustment of the NCP is desirable, provided the information on which the adjustment is based is as accurate and precise as the information used in the valuations done every four years, as now required under PL 98–94.

The value of changing to an accrual method that allocates a higher charge to personnel at higher years of service, nearing 20 years,

[1] As a refinement, the DoD Actuary could evaluate whether longevity currently differs and is expected to differ in the future, and use this information in determining service-specific accrual charges.

has no additional value to the current system from a long-term perspective. However, it is valuable to a service in the short term only if it can obtain immediate credit for the decrease in retirement liability from separating senior personnel, or pay the immediate charge from keeping additional senior personnel. But again, the service does not have a property right to the cost decrease or a source of revenue to pay the increase, and as under the current system the service interacts with Congress. Furthermore, accounting conventions, such as entry-age normal or Dahlman's approach, for example, do not provide estimates of actual cost but only of accounting charges, and these are pass-throughs to Treasury. The suggestion we make for any policy change affecting the retirement liability is to do a separate analysis to estimate the change in the liability and include this information in the policy discussion. Policymakers can use this information in deciding whether and how much to change other line items in a service's budget.

If we turn attention from these issues and focus on the accrual charge, past studies and our research make a case for changing to service-specific NCPs, and, within a service, on NCPs for officers and enlisted personnel. This would provide accurate information about total and marginal accrual costs, unlike the current system. It should also eliminate controversy over whether a service is being overcharged and is cross-subsidizing another service. Thus, the change should bring more light and less heat.

The shift to service-specific NCPs does not seem too difficult to accomplish. First, the DoD Actuary will need time and resources to develop the capability to compute the NCPs. Much of this will be a replication of the existing system, but there are always details that take time, e.g., how to handle a service member who starts in one service and moves to another, or how to handle the flow of personnel from a given AC service to any of the six Selected Reserve components. The Actuary will need to prepare a plan and cost estimate of what is required to create the new capability.

Second, DoD and Congress will need to come to an agreement about how to adjust service budgets. An approach that would make no service worse off than today would be to assure each service that its new accrual charge would be fully covered and that other parts of its budget would be unchanged. But given the dynamic nature of budgeting, with budgets evolving to meet the needs of current operations and respond to strategic assessments, other parts of the budget would not be held constant. Perhaps a more-realistic view would be to continue the current approach in which Congress works with the military services to create budgets that cover anticipated requirements, including personnel costs and accrual costs. There would be no discussion of retrieving past overfunding or charging for past underfunding. The services, DoD, and Congress would continue to interact in producing the annual defense authorization and appropriation acts, and service-specific NCPs, by providing more accurate NCPs to better inform this process and lead to different resource allocations. We have not attempted to determine that extent to which the current system has caused inappropriate resource allocation, nor to estimate the extent to which a service-specific system would improve resource allocation. But we do observe that a service-specific system should weaken, if not eliminate, service concerns that the single NCP leads to one service benefiting at the expense of other services.

Finally, our DRM-based estimates find a significant inaccuracy in the accrual charge running into the hundreds of millions of dollars for a service. If it were a small disparity, little would be gained from service-specific NCPs. With a large inaccuracy, though, the accrual charge would be a continuing and legitimate annoyance to a service facing top-line budget guidance and wanting to program resources for uses other than retirement costs. We showed that the inaccuracy exists under the current retirement system, as well as under an alternative system of the type now being considered for retirement reform.

APPENDIX

Inaccuracy in Cost Changes from an Increase in Basic Pay

This appendix considers the change in cost from an increase in basic pay when the accrual charge is based on a single NCP versus a service-specific NCP. Two costs are considered: the total cost of an entering cohort, represented by the present discounted value of basic pay over the cohort's service life plus the accrual cost; and the accrual cost alone. The reason for looking at the accrual cost alone is that in some cases, budgeting focuses on the accrual cost.

Service-specific NCP Case

Total cost: $W_1(1 + \theta_1)$.

Change in Total Cost with Respect to an Increase in Basic Pay for Service One

The increase in basic pay is across the board and results in a direct increase in the present discounted value of basic pay for an entering cohort over its years of military service and an indirect increase resulting from the effect of higher pay on the retention rate.

$$\frac{dW_1(1+\theta_1)}{db} = \frac{\partial W_1(1+\theta_1)}{\partial b} + \frac{\partial W_1(1+\theta_1)}{\partial r}\frac{\partial r}{\partial b}$$

$$\frac{\partial W_1(1+\theta_1)}{\partial b} = W_{1b}(1+\theta_1) + W_1\theta_{1b}$$

$$\frac{\partial W_1(1+\theta_1)}{\partial} = W_{1r}(1+\theta_1) + W_1\theta_{1r}.$$

Since $\theta_1 = L_1 / W_1$,

$$\theta_{1b} = \frac{L_{1b} - \theta_1 W_{1b}}{W_1}$$

$$\theta_{1r} = \frac{L_{1r} - \theta_1 W_{1r}}{W_1}.$$

Substituting

$$\frac{\partial W_1(1+\theta_1)}{\partial b} = W_{1b}(1+\theta_1) + L_{1b} - \theta_1 W_{1b} = W_{1b} + L_{1b}$$

$$\frac{\partial W_1(1+\theta_1)}{\partial r} = W_{1r}(1+\theta_1) + L_{1r} - \theta_1 W_{1r} = W_{1r} + L_{1r}.$$

Therefore,

$$\frac{dW_1(1+\theta_1)}{db} = W_{1b} + L_{1b} + (W_{1r} + L_{1r})\frac{\partial r}{\partial b}.$$

Change in Accrual Cost with Respect to an Increase in Basic Pay for Service One

This is a portion of the analysis above. Accrual cost is $W_1\theta_1$, and the change in accrual cost with respect to a change in basic pay is

$$\frac{dW_1\theta_1}{db} = \frac{\partial W_1\theta_1}{\partial b} + \frac{\partial W_1\theta_1}{\partial r}\frac{\partial r}{\partial b}$$

$$\frac{\partial W_1\theta_1}{\partial b} = W_{1b}\theta_1 + W_1\theta_{1b} = W_{1b}\theta_1 + (L_{1b} - \theta_1 W_{1b}) = L_{1b}$$

$$\frac{\partial W_1\theta_1}{\partial r} = W_{1r}\theta_1 + W_1\theta_{1r} = W_{1r}\theta_1 + (L_{1r} - \theta_1 W_{1r}) = L_{1r} ,$$

where expressions for θ_{1b} and θ_{1r} from above are substituted after the second equal sign. As seen, the change in accrual cost operates through the change in the retirement liability and is

$$\frac{dW_1\theta_1}{db} = L_{1b} + L_{1r}\frac{\partial r}{\partial b} .$$

Single NCP Case

Total cost: $W_1(1 + \theta)$.

Change in Cost with Respect to an Increase in Basic Pay for Service One

The increase in basic pay is across the board and results in a direct increase in the present discounted value of basic pay for an entering cohort over its years of military service and an indirect increase resulting from the effect of higher pay on the retention rate:

$$\frac{dW_1(1+\theta)}{db} = \frac{\partial W_1(1+\theta)}{\partial b} + \frac{\partial W_1(1+\theta)}{\partial r}\frac{\partial r}{\partial b}$$

$$\frac{\partial W_1(1+\theta)}{\partial b} = W_{1b}(1+\theta) + W_1\theta_b$$

$$\frac{\partial W_1(1+\theta)}{\partial r} = W_{1r}(1+\theta) + W_1\theta_r .$$

Since $\theta = (L_1 + L_2)/(W_1 + W_2)$,

$$\theta_b = \frac{L_{1b} - \theta W_{1b}}{W_1 + W_2}$$
$$\theta_r = \frac{L_{1r} - \theta W_{1r}}{W_1 + W_2} .$$

Substituting and letting $\alpha \equiv W_1/(W_1 + W_2)$,

$$\frac{\partial W_1(1+\theta)}{\partial b} = W_{1b}(1+\theta) + \alpha(L_{1b} - \theta W_{1b}) = W_{1b} + L_{1b} + (1-\alpha)(\theta W_{1b} - L_{1b})$$
$$\frac{\partial W_1(1+\theta)}{\partial r} = W_{1r}(1+\theta) + \alpha(L_{1r} - \theta W_{1r}) = W_{1r} + L_{1r} + (1-\alpha)(\theta W_{1r} - L_{1r}) .$$

Therefore,

$$\frac{dW_1(1+\theta)}{db} = W_{1b} + L_{1b} + (1-\alpha)(\theta W_{1b} - L_{1b}) + [W_{1r} + L_{1r} + (1-\alpha)(\theta W_{1r} - L_{1r})]\frac{\partial r}{\partial b} .$$

Grouping terms and using the result for the service-specific NCP case,

$$\frac{dW_1(1+\theta)}{db} = \frac{dW_1(1+\theta_1)}{db} + (1-\alpha)[(\theta W_{1b} - L_{1b}) + (\theta W_{1r} - L_{1r})\frac{\partial r}{\partial b} .$$

This implies that when a single NCP is used, the inaccuracy in the change in total cost from a service-specific across-the-board increase in basic pay depends on the sign of the second term on the right-hand side. The sign of the term is theoretically indeterminate, so this is an empirical question.

Change in Accrual Cost with Respect to an Increase in Basic Pay for Service One

This is again a portion of the analysis above. Accrual cost is $W_1\theta$, and the change in accrual cost with respect to a change in basic pay is

$$\frac{dW_1\theta}{db} = \frac{\partial W_1\theta}{\partial b} + \frac{\partial W_1\theta}{dr}\frac{\partial r}{\partial b}$$

$$\frac{\partial W_1\theta}{\partial b} = W_{1b}\theta + W_1\theta_b = W_{1b}\theta + \alpha(L_{1b} - \theta W_{1b}) = L_{1b} + (1-\alpha)(W_{1b}\theta - L_{1b})$$

$$\frac{\partial W_1\theta}{\partial r} = W_{1r}\theta + W_1\theta_r = W_{1r}\theta + \alpha(L_{1r} - \theta W_{1r}) = L_{1r} + (1-\alpha)(W_{1r}\theta - L_{1r}),$$

where expressions for θ_{1b} and θ_{1r} from above are substituted after the second equal sign. In contrast to the result for the service-specific NCP case, the change in accrual cost now operates through both the change in the retirement liability and the change in the wage bill:

$$\frac{dW_1\theta}{db} = L_{1b} + L_{1r}\frac{\partial r}{\partial b} + (1-\alpha)(W_{1b}\theta - L_{1b}) + (1-\alpha)(W_{1r}\theta - L_{1r})\frac{\partial r}{\partial b}$$

$$\frac{dW_1\theta}{db} = \frac{dW_1\theta_1}{db} + (1-\alpha)[(W_{1b}\theta - L_{1b}) + (W_{1r}\theta - L_{1r})\frac{\partial r}{\partial b}].$$

Comparing this with the result for the change in the full cost, we see that the bias terms are the same.

Abbreviations

AC	active component
CBO	Congressional Budget Office
CRDP	Concurrent Retirement and Disability Pay
DoD	Department of Defense
DRM	dynamic retention model
FY	fiscal year
MCRMC	Military Compensation and Retirement Modernization Commission
MRF	Military Retirement Fund
NCP	normal cost percentage
PAYGO	pay-as-you-go
PL	Public Law
QRMC	Quadrennial Review of Military Compensation
RC	reserve component
SMCR	Standard Military Composite Rate
TSP	Thrift Savings Plan
VA	Department of Veterans Affairs
VSI	Voluntary Separation Incentive
WEX	Work Experience File

References

Asch, Beth J., James Hosek, and Michael G. Mattock, *A Policy Analysis of Reserve Retirement Reform*, Santa Monica, Calif.: RAND Corporation, MG-378-OSD, 2013. As of January 25, 2016:
http://www.rand.org/pubs/monographs/MG378

———, *Toward Meaningful Compensation Reform: Research in Support of DoD's Review*, Santa Monica, Calif.: RAND Corporation, RR-501-OSD, 2014. As of July 23, 2015:
http://www.rand.org/pubs/research_reports/RR501.html

Asch, Beth J., Michael G. Mattock, and James Hosek, *Reforming Military Retirement: Analysis in Support of the Military Compensation and Retirement Modernization Commission*, Santa Monica, Calif.: RAND Corporation, RR-1022-MCRMC, 2015. As of January 25, 2016:
http://www.rand.org/pubs/research_reports/RR1022.html

Committee on Armed Services, House Report No. 98-107, accompanying House Resolution 2969, 98th Congress, first session, 1983. As of January 29, 2016:
https://www.congress.gov/bill/98th-congress/house-bill/2969

Congressional Budget Office, "Accrual Accounting for Military Retirement: Alternative Approaches," Staff Working Paper, July 1983.

———, "Assessing Pay and Benefits for Military Personnel," August 15, 2007, p. 3. As of January 29, 2016:
http://www.cbo.gov/sites/default/files/cbofiles/ftpdocs/85xx/doc8550/08-15-militarycompensation_brief.pdf

Dahlman, Carl J., *The Cost of a Military Person-Year: A Method for Computing Savings from Force Reductions*, Santa Monica, Calif.: RAND Corporation, MG-598-OSD, 2007. As of January 26, 2016:
http://www.rand.org/pubs/monographs/MG598.html

DoD Actuary—*See* U.S. Department of Defense Office of the Actuary

Eisenman, Richard L., David W. Grissmer, James Hosek, and William W. Taylor, *The Accrual Method for Funding Military Retirement*, Santa Monica, Calif.: RAND Corporation, MR-811-OSD, 2001. As of January 26, 2016: http://www.rand.org/pubs/monograph_reports/MR811.html

Gotz, Glenn A., *Military Retirement System Costing and Budgeting*, RAND Corporation, unpublished working draft, May 1985.

Hix, William M., and William W. Taylor, *A Policymaker's Guide to Accrual Funding of Military Retirement*, Santa Monica, Calif.: RAND Corporation, MR-760-A, 1997. As of January 26, 2016: http://www.rand.org/pubs/monograph_reports/MR760.html

Hogan, Paul F., and David K. Horne, *Military Retirement Accrual Charge as a Signal for Defense Resource Allocation*, Alexandria, Va.: United States Army Research Institute for the Behavioral and Social Sciences, December 1989.

Mattock, Michael G., Beth J. Asch, and James Hosek, *Making the Reserve Retirement System Similar to the Active System: Retention and Cost Estimates*, Santa Monica, Calif.: RAND Corporation, RR-530-A, 2014.

MCRMC—*See* Military Compensation and Retirement Modernization Commission

Military Compensation and Retirement Modernization Commission, *Concepts for Modernizing Military Retirement*, white paper, Washington, D.C., March 2014.

———, *Report of the Military Compensation and Retirement Modernization Commission, Final Report*, Washington D.C., January 2015. As of July 23, 2015: http://www.mcrmc.gov/public/docs/report/MCRMC-FinalReport-29JAN15-LO.pdf

U.S. Department of Defense Office of the Actuary, *Valuation of the Military Retirement System: September 30, 2011*, Washington, D.C., February 2013. As of January 29, 2016: http://actuary.defense.gov/Portals/15/Documents/MRFValRpt2011.pdf

———, "Actuarial Work for the Chief Financial Officers Act Financial Statements," memorandum to file, Washington, D.C., March 2014a. As of January 29, 2016: http://actuary.defense.gov/Portals/15/Documents/CFO-financial-statements.pdf

———, *Valuation of the Military Retirement System: September 30, 2012*, Washington, D.C., April 2014b. As of January 29, 2016: http://actuary.defense.gov/Portals/15/Documents/MRF_ValRpt2_2012.pdf